Catecholamines in Normal and Abnormal Cardiac Function

Advances in Cardiology

Vol. 30

Series Editor
J.J. Kellermann, Tel Hashomer

S. Karger · Basel · München · Paris · London · New York · Sydney

Catecholamines in Normal and Abnormal Cardiac Function

William Muir Manger, MD, PhD, FACP, FACC
Associate Professor of Clinical Medicine,
Department of Medicine, New York University Medical Center and
Institute of Rehabilitation Medicine, Lecturer in Medicine, College of Physicians
and Surgeons, Chairman National Hypertension Association, Inc., New York,
New York

37 figures and 19 tables, 1982

S. Karger · Basel · München · Paris · London · New York · Sydney

Advances in Cardiology

National Library of Medicine, Cataloging in Publication

Manger, William Muir, 1920–
Catecholamines in normal and abnormal cardiac function/William Muir Manger.
– Basel; New York: Karger, 1982.
(Advances in cardiology; v. 30)
1. Catecholamines – pharmacodynamics 2. Heart – drug effects 3. Heart Function Tests
I. Title II. Series
W1 AD53C v. 30 [WK 725 M277c]

ISBN 3–8055–3516–3

Contents

Contents

Foreword

When I started my career in cardiovascular research in 1947, *von Euler* had just demonstrated from his classic studies on the splenic nerve of the ox, that norepinephrine was the sympathetic neurotransmitter. The years since have seen a remarkable unfolding of the extraordinary complex events at the sympathetic neuroeffector junctions in the heart and blood vessels, whose outcome dictates the ability of the cardiovascular system to respond appropriately to the many stresses to which the body is subjected. Indeed so much information is now available of the factors controlling the release of norepinephrine, its action on alpha$_1$ and alpha$_2$ receptors and its metabolism that it is difficult for those not active in the field to encompass and comprehend the current state of knowledge.

Dr. *Manger's* monograph meets this challenge admirably as befitting his long years of study and contributions to this field. I am proud that my own institution can claim to have stimulated his early interest in the catecholamines. At the time that Dr. *Manger* commenced his fellowship in the Mayo Graduate School of Medicine, the Mayo Clinic was gaining increasing familiarity with the diagnosis and treatment of patients with pheochromocytoma. Working in the laboratories of Drs. *Bollmann*, *Wakim* and *Baldes*, Dr. *Manger* developed, in 1953, a technique for the chemical quantitation of epinephrine and norepinephrine in the plasma of patients. In 1977 in association with another colleague from his Mayo years, Dr. *Ray Gifford*, their classic monograph on pheochromocytoma was published. This provides the most complete account of the morphology, pharmacology and the clinical aspects of this remarkable tumor, and is the hallmark of their broad experience and leadership.

Apart from his own research, *Bill Manger's* enthusiasm, his ability to communicate effectively with colleagues and his friendly personality have served to stimulate interest worldwide in the sympathomimetic amines. His associations with the most illustrious names in the field, including *Julius Axelrod* and *Ulf von Euler*, led to the founding of the Catecholamine Club in 1968. As the secretary-treasurer from the beginning Dr. *Manger* has been the prime mover in spearheading the meetings which have set the standard for excellence and served so well to advance the science. This year, as a proper recognition of his broad contributions he has been elected President. The National Hypertension Association Inc. which he started in 1977 with its prominent members of its Board of Trustees and its distinguished International Medical Advisory Council serves to unite physicians, scientists and lay public in the continued quest for the better understanding of high blood pressure. This is complimentary to his early interests, since the sympathetic nervous system has the key role in the regulation of blood pressure in normal circumstances, and it is evident that it plays a significant or dominant role in many disorders of circulatory function.

The present book is characterized by lucid and succinct writing, and clear illustrations. His familiarity with the literature has permitted him to focus objectively on the key advances in our knowledge of the role of the adrenergic nervous system and of circulatory catecholamines in governing the performance of the heart, both in normal and abnormal states. It will serve as a springboard for those young in years from which to formulate new ideas, and put them to the test, and it will serve as a reference for all of us whose memories of particular experiments and concepts need refreshing.

John T. Shepherd, MD, DSc, DSc (Hons) FRCP,
Director for Education, Mayo Clinic and Foundation
Dean Mayo Medical School

Dedication

Compassion, concern, sympathy and sensitivity are qualities which are deeply admired by all. The three men to whom this book is dedicated have displayed these qualities to the full limit.

Their care for students, associates and their fellow men has been a unique example and remarkable inspiration which will remain always in the memory of those fortunate enough to have known them.

'Almost all noble attributes – courage, love, hope, faith... loyalty – can be transmuted into ruthlessness. Compassion alone stands apart from the continuous traffic between good and evil... within us. Compassion is the antitoxin of the soul' *(Eric Hoffer)*.

There can be no improvement on the statement by *Albert Schweitzer* that: 'Example is not the most important thing in influencing others; it is the *only* thing.'

It has been said that the essence of Christianity is: 'To see a need and fill it.' This credo has been central in the lives of Canon *Charles Martin*, Dr. *Howard Rusk* and Dr. *Shields Warren*.

William Muir Manger

Acknowledgements

The extraordinarily fine assistance of my research associate, *Mildred Hulse*, in preparing, editing, and significantly improving this manuscript was invaluable. The assistance of *Richard Seides, Richard Sussman, Kirwan Webb, Thomas Brown* and Reverend *Don Bundy* in proofreading, referencing, and editing was also extremely helpful. I am particularly grateful to Drs. *Howard B. Burchell, Brian F. Hoffman, Michael P. Kaye, Robert J. Lefkowitz, Raymond Pruitt, Michael R. Rosen, John T. Shepherd* for their helpful comments and constructive suggestions regarding the manuscript.

This review was supported by the National Hypertension Association, Inc., the Pew Memorial Trust, the Hypertension Fund of the Institute of Rehabilitation Medicine, the Hess Foundation, the Hearst Foundation, and the Northrup Foundation.

Canon Charles Martin

Headmaster and Teacher, 1906

There is no more important endeavor than the molding of the character of youth; these young people are the 'wave of the future' and the course of human events depends on the direction they take. As Headmaster of Saint Albans School, Canon *Charles Martin* established an unparalleled standard of excellence in his students. To instill worthy ideals and spiritual guidance permits noble dreams which shape the future. By his example, remarkable in every area, *Charlie Martin* has contributed so much to the lives of so many. For of the most high cometh teaching.

William Muir Manger

Mr. *Alfred R. True*, formerly Assistant Headmaster and Head of the Lower School at Saint Albans and a modern 'Mr. *Chips*' in the fullest sense, has been an intimate associate and close friend of *Charlie Martin* for 33 years. His deep admiration and warm affection for *Charlie Martin* is expressed in the following:

'Canon *Charles Martin* has an open and refreshing interest in people and events, and he cares deeply about both. His concern for others and his sensitive involvement in their welfare underlie and explain his 28 years of outstanding success as headmaster of Saint Albans School, where he was greatly beloved, and where he made everyone feel needed in a close school family. Now engaged in other work, he continues to be the cherished friend of all who know him and who count on hearing his warm and reassuring voice.

'Whenever his help is sought or needed, he finds no hour too inconvenient, no day too filled with conflicting obligations, no road to be traveled too long, no task too difficult, in responding. He has given help and solace to countless numbers on countless occasions.

'He has a searching and inquiring mind which leads him to explore and examine all sides of a question, but his ready sense of humor keeps things in perspective. His boundless, restless energy fortifies his widely diverse efforts and sustains his vigorous pursuit of goals. He sets high standards for himself and expects no less of others. In his view, no job is so well performed that it can't be improved. He reaches for better performance himself and believes firmly that others should be encouraged and pressured to put possible dormant and unchallenged talents to use. His infectious enthusiasm in approaching those whose ability he senses and whose help and participation he wants, overcomes resistance and doubt. Because he is engagingly adept in the art of persuasion, many of his friends and colleagues have learned the joy and satisfaction of succeeding in assignments they were initially reluctant to accept. He has the great gift of stimulating others to useful action, "to their own ultimate benefit", he says, and he has the satisfaction of knowing that he has brightened the days and lightened the burdens of many.'

Mr. *John C. Davis*, Assistant Headmaster and Head of the Upper School at Saint Albans and a remarkable educator himself, has also been an intimate associate and close friend of *Charlie Martin* for 33 years. His following remarks reveal something about the nobility of character and the brilliance of leadership which have established Canon *Martin's* reputation as one of the great headmasters of our time.

'Among the glories of American private education – and it is a record of such that goes back to the seventeenth century – is the continuing presence of charismatic, dedicated, and innovative individuals.

'Innovative in their desire to propel education from the springboard of the present, dedicated to the mysterious amalgams of the ideals of both this world and the next, and charismatic in their ability to attract, compel, cajole, or bewitch with these ideals the American student of 13–21 years of age, such men and women have been in themselves the history of American education.

'*Charles Samuel Martin*, Headmaster of St. Albans School in Washington, from 1949 to 1977, is one of these. Born in Philadelphia – that city of schools of quiet distinction – he taught at Episcopal Academy and served for seven years as parish priest in Vermont before embarking on his larger parish of St. Albans School in 1949.

'For a headmaster of a school like St. Albans, this swarming parish of boys and their families does nothing but increase and multiply. Not only did the School's size grow from 364 to 530, but wives, children, and friends were added to the number who came increasingly for advice, help, information, consolation, or simple handholding. Nor was anyone turned away, and a some-time secretary ruefully looked at the dictation that had magically expanded over a holiday weekend and said: "He's the only man I know who ever answered a thank-you note."

'It was by this network of ghostly counsel and support that, both unconsciously and by design, he encouraged his ideals of education. Starting with the practical problems at hand – the academic problem of the student, the marital difficulties of the parents, the loss of the disconsolate, he moved into the spiritual world of character formation, reconciliation, and compassion. He was always aware of the priority of God's jobs to those of Mammon, and if Mammon's were demanding too much time, so much the worse for them. "To the mischief with it!" was his harsh judgement upon them.

'*Charlie Martin* was not an educational theorist. Although in his time he saw St. Albans among the first to introduce Russian and outdoor survival programs, penology and African studies, social service in the ghetto, or using the media as the educational message, he was eminently pragmatic. He wasted little time on talk, moved at once to action, and once, when he was over 60, he rappelled down a forty-foot wall to show worried mothers that rock-climbing was a safe activity. (They were not convinced.) And when his harried administrators thought they had completed the job at hand and had a program working, there was *Charlie Martin* off stirring up educational

ant hills in another area and wanting to send students to work in the back
country of Tanzania or in the Richmond penitentiary. There was no rest for
anybody, most of all for him, nor was rest considered desirable.

'Long before politicians had rediscovered the work ethic, *Charles
Martin* was preaching it. In the midst of the late-60s' youth revolution, he
grumbled publically about "what they thought was wrong with the Puritan
work ethic." In those troubled times when deans and heads of academic
institutions were being held hostage in their own offices, *Charles Martin* was
busily anticipating their demands, organizing their reasonable educational
expectations into a program, and in essence – keeping miles ahead of them.
Imaginative foresight was one of his great gifts.

'His administration at St. Albans was marked by the development of
the school structure, the enlargement of the concept of the school community
and its social responsibility, and his awareness of the increased role non-
academic education should play in the broad practice of instruction. Like the
innovative *Eliot* of Harvard, Dr. *Nott* at Union, and *Hutchins* at Chicago,
he saw that the present should not be enshrined eternally in an unchanging
carapace, for the moment became rapidly the educational past. For him the
future of education was like the old comic strip of the frankfurter tied to a
stick, in turn tied to the dog's back, so that the animal chased the future
hotdog forever. Restless, never satisfied, creating new goals as he pursued
his high ideals, *Charles Martin* made a good school great, and a great ideal
more capable of realization.

'This he could do because, at the bedrock of his character, he considered
the humanity of man important, but not so important as its divine ingredient.
When he said once that: "He was not trying to get students into Harvard but
into the Kingdom of Heaven," and a bright lad muttered audibly "when is
the Director of Admissions going to visit?" *Charles Martin* only smiled.
Perhaps he knew.'

Dr. Howard A. Rusk

Father of Rehabilitation Medicine and Teacher, 1901

Few individuals in a profession have contributed a greater measure to mankind than Dr. *Howard Rusk*. Through his efforts in rehabilitation medicine, often assisted by his charming and very talented wife, the late *Gladys H. Rusk*, his influence has been worldwide. The dedication to his work and the devotion to his cause – to understand the handicapped and to help the disabled help themselves – have endeared him to the hearts of all nations. In a true sense he is a champion and a hero of compassion and peace.

William Muir Manger

In the following, Dr. *Irvine H. Page*, an internationally distinguished scientist, has expressed his sense of gratitude and tribute to his long time friend, Dr. *Howard A. Rusk:*

A Note of Quiet Gratitude

'*Howard Rusk* is a remarkable fellow. I first heard of him while I was serving on the clinical research committee of the National Research Council during World War II. Repeatedly, *Howard Rusk* was refered to as an innovator in a field that few thought could be "innovated", i.e. rehabilitation, otherwise known as "physical therapy". It was considered a drab subject. But the Air Force seemed to think differently, chiefly because of this young man, *Howard Rusk*, a private practitioner from St. Louis. History tells all the rest.

'*Howard* combines in a remarkable fashion many talents, chief of which are knowledge, ability, compassion and charm. He is a first-rate doctor, a pretty good research worker, and a topnotch speaker and organizer. He sought the ear of everyone to further his unselfish objective, and got it. Much is made of his friendship with celebrities, but I can testify after having known him for 30 years that his personal warmth is given to all who approach him. Unlike many, he has created and developed what he promised. Although he has been showered with honors, his life will have left a more permanent heritage in the form of rehabilitation of the handicapped. His past achievement would be enough for most of us but *Howard* will never give up and rest on his laurels. For that we all bless him.'

Dr. Shields Warren

Pathologist and Teacher, 1898–1980

Dr. *Shields Warren* was revered and loved by all who knew him. His scientific brilliance combined exceptional ingenuity and boundless energy with rigid objectivity and inflexible integrity. His charm, humility and warmth made it pure delight to be in his presence; his concern and compassion for his fellowmen was no less than magnificent. He was a man for all seasons. A statement made by Sir *Dominic J. Corrigan* in 1829 so perfectly describes the life-long attitude of *Shields Warren* that it is included in this dedication:

'Whether my observations and opinions be disproved or supported, I shall be equally satisfied. Truth is the prize... and, in the contest, there is at least this consolation, that all the competitors may share equally the good attained.'

William Muir Manger

The following tribute to Dr. *Shields Warren* was expressed by Dr. *John Z. Bowers*, a close friend and former research associate.

Shields Warren – A Tribute

'Few men have had as deep and varied an influence on American medical science as *Shields Warren*. He can be described as a "universal genius". Yet at the same time he was a remarkably kind, gentle, and self-effacing man, everyone's friend, and without an enemy. Those of us who enjoyed a personal as well as professional association with *Shields* admired and benefitted from his infinite capacity for hard work.

'The fact that his grandfather, *William Fairfield Warren*, a Methodist cleric, was the founding president of Boston University, and his father, *William Marshall Warren*, served for many years as professor and dean of the College of Liberal Arts, influenced *Shields* to enroll at the university for his undergraduate studies. After graduation in 1918 and a year of "hoboing" across America, *Shields* studied medicine at mighty Harvard. In that period all students were required to conduct and report a sanitary survey. *Shields* selected Rochester, New Hampshire, and had as his collaborator a fellow student at Boston University, *Alice Springfield*, from the same community. The selection of Rochester introduced a romantic note for not long thereafter she became *Alice Springfield Warren*, his charming wife, helpmate, and the mother of two lovely daughters.

'After training in pathology, then the queen of the basic sciences, with *Frank Burr Mallory* at Boston City Hospital, *Shields* joined the Harvard faculty and 2 years later became head of pathology at the "Deac", the New England Deaconess Hospital. He worked closely with *Elliot P. Joslin*, deemed by many the world's leading expert on diabetes mellitus. This association resulted in *Shields*' magisterial book, *The Pathology of Diabetes Mellitus*. Thyroid tumors became another of his areas of expertise when *Shields* took over Surgical Pathology for the Lahey Clinic. The clinic was dominated by excellent surgeons led by *Frank Lahey* and *Richard Cattell*. (Perhaps *Cattell's* greatest hour came when he was lecturing in London and was called in consultation on *Anthony Eden* who had suffered a torn common bile duct during biliary surgery. *Cattell* had been successful in repairing a number of such cases, and he agreed to operate on *Eden* but stipulated that this could only be done at a New England hospital. Rumor has it that *Winston Churchill* tried to persuade *Cattell* to perform the surgery in London, but *Cattell* insisted on Boston, and became victorious.)

'In the late 1920s *Shields* began his pioneering and classical research on the effects of ionizing radiation on living tissues. This culminated in his selection as Commander *Warren* to lead a navy team to Nagasaki in September 1945 for elucidation of the medical effects of the atomic bomb. Influenced by his months at Nagasaki, *Shields* became a key figure in the creation of the Atomic Bomb Casualty Commission for long-term studies of the effects of the weapon. In 1946 he prepared a statement proposing such a commission; it was drafted by Navy Secretary *James Forrestal* and submitted to President *Harry S. Truman*. With *Truman's* prompt endorsement, the ABCC came into full flower in the spring of 1947 while *Shields* was renewing his studies at Nagasaki. He continued as its advisor and by invitation returned to Japan on a final visit in 1975 when the ABCC became the Radiation Effects Research Foundation under Japanese leadership but with financial support from both governments.

'*Shields* never retired from active service; he continued to be involved in governmental medicine and radiation, international studies, and directorships of such major companies as Mallinckrodt. The walls of his office in the Shields Warren Radiation Laboratory at Harvard were covered with pictures of his disciples and a world map adorned with innumerable flags representing the homes of his devoted students.

'He loved the sea and was an ardent and excellent sailor. A "radioactive flag" showing an atomic nucleus surrounded by electron shells flew from the mast of his boat. *Shields* was truly a man for all seasons, mourned and beloved by hosts of people around the world.'

Dr. *Bowers* concluded a eulogy to Dr. *Warren* with the following:

'*Shields* was constantly on the lookout for new vistas, new challenges, and new intellectual horizons. He was the epitome of the intellectual explorer. As *Rudyard Kipling* wrote in his poem, *The Explorer:*

"– On one everlasting Whisper day and night repeated – so:
Something hidden. Go and find it. Go and look behind the Ranges –
Something lost behind the Ranges. Lost and waiting for you. Go!" '

Author's Remark

Knowledge of the importance of the adrenergic system in normal and abnormal cardiac function has expanded rapidly and remarkably within the past two decades. It therefore seemed justifiable to review and concisely compile this information into one volume so that it is readily accessible.

This monograph briefly presents our understanding of catecholamine metabolism, adrenergic receptors and responses, and neural regulation of the heart. It also describes some of the current concepts of involvement of the adrenergic system and circulating catecholamines in cardiac pathophysiology.

William Muir Manger

Introduction

In his textbook on medical anthropology from the end of the 18th century, *J.C. Loder* comments upon the possible function of the suprarenals in the following words (translated from the Swedish edition, Lund, 1799): 'In the adult they may contribute to give the blood in the lower caval veins some sharpness, in order better to stimulate the heart to contraction...'. That this surmise was closer to the truth than the author could anticipate was proven some 100 years later by the dramatic effects of suprarenal extracts on cardiovascular activity, observed by *Oliver and Schäfer* in 1895. This discovery made instantly clear that the heart could be subject to the potent regulatory effects of certain naturally occurring agents, later identified as catecholamines.

The accelerating effect of sympathetic heart nerve stimulation demonstrated some 20 years earlier by *M. and E. von Cyon*, and at about the same time by *Bever and von Bezold*, hardly evoked the interest it deserved, in part probably due to lack of understanding of the mechanism governing this effect.

At the turn of the century it was thus known that the heart could be excited by nerve stimulation as well as by suprarenal extracts. An integration of these two effects was achieved when *Elliott* in 1904 proposed his famous hypothesis that the chromaffin cell hormone adrenaline served as chemical neurotransmitter. This was followed by the genially simple experiments of *Loewi* in 1921 providing the final solution to a problem which had occupied physiologists for so long.

Continued experimental work along these lines revealed that adrenaline was not the only member of the catecholamine family endowed with powerful regulatory properties in the heart, but had to share this important action with the primary amine dopamine, and its beta-hydroxylated congener, noradrenaline.

In the past decades the role of the catecholamines for heart function in normal and diseased states has become the subject of intensive studies, to a large part based on new and improved techniques, both with regard to evaluation of the manifold catecholamine effects on the heart and with regard to the formation and metabolism of the active amines. These problems are extensively dealt with by the author, whose own interest in and contributions to this field extend over several decades. In a systematic and engaging way the author presents and discusses the main areas in which the catecholamines influence cardiac function.

The recognition of specific receptors in the heart, and the discovery of numerous agonists and antagonists, some of which possess a high degree of specificity, has made possible precision and effectiveness in the treatment of disturbances in heart function which in their turn have become increasingly better understood and identified. This applies to coronary circulation as well as to the mechanical and metabolic processes in the myocardium. The mode of action of a variety of cardiac drugs, in later years elucidated at chemical and physiochemical levels, are exemplified by the mechanisms of action of adenylate cyclase-stimulating and catecholamine-depleting drugs. The unique arrangement of the contractile tissue in the heart and its nervous control offer special conditions for intervention by the highly active catecholamines. Still many unanswered challenging questions and problems remain as regards the fine regulation and reflex control of cardiac function.

The rapid and impressive advances in diagnosis and treatment of cardiac dysfunctions during recent years have increased the needs for textbooks covering this multifaceted field. The present volume gives a highly readable and critical evaluation of the role of the catecholamines in heart function, not only for the benefit of those who have the responsibility of giving a scientifically based treatment to the heart patient, but also for those more directly engaged in cardiological research. The logical pursuit of the numerous manifestations of catecholamine actions in the heart allows the reader to obtain an integrated view of the present knowledge in this central sector of medicine.

Ulf Svante von Euler, Stockholm
Emeritus Professor of Physiology
Karolinska Institute
1970 Nobel Prize Laureate in Physiology

Preface

In this book Dr. *Manger* has drawn upon his longstanding interest in the effects of catecholamines on the heart and circulation to provide a remarkably complete and clear summary of our current understanding of the biochemistry, physiology and pharmacology of adrenergic transmitters, the autonomic regulation of cardiac electrical and mechanical activity and the role of adrenergic transmitters in causing cardiovascular pathology. The text is admirably illustrated and throughout the author has provided clear and meaningful tables to summarize the most important information and relationships.

The short historical background highlights a number of the major findings during the past 60 years and emphasizes the manner in which understanding of biological processes and systems grows in parallel with technological advances. The next section provides a lucid exposition of the current understanding of the synthesis, storage and release of catecholamines in nerve terminals and chromaffin cells, a detailed consideration of the regulation of these processes and a clear picture of the relative importance of the different mechanisms in relation to physiological control and drug action. This section is followed by a detailed consideration of adrenergic receptors and the responses that result from their activation. The introductory material summarizes in tables the typical responses of most tissues to activation of α- and β-receptors, the commonly employed α- and β-agonists and antagonists and finally the effects of circulating epinephrine and norepinephrine. The reader then is given a concise description of current knowledge of the types of adrenergic receptors in the myocardium and coronary circulation and the effects of their activation. For each case evidence is presented fully and interpreted fairly and an adequate number of appropriate references is provided to permit the reader to clarify points of particular interest.

This background permits a detailed description of the neural regulation of the heart. The interactions between sympathetic and vagal effects are described as are the major cardiac reflexes and the role of intracranial and spinal neural mechanisms. A consideration of the effects of neural mediators on cardiac electrical activity and the coronary circulation and the changes that result from denervation then permits a full exposition of the known and expected effects of autonomic influences on the ECG and cardiac rhythm. This material benefits from the author's understanding of both basic physiology and clinical cardiology.

The final section considers the important involvement of the autonomic nervous system in cardiac pathology and pathophysiology. The problem of infarction is given prominence as is the role of the adrenergic nervous system in ischemia and infarction and the associated disturbances of rhythm. There are interesting treatments of the changes in adrenergic physiology in heart failure and the role of the sympathetic nerves in causing hypertrophy. The final topic, a consideration of the effects of excessive adrenergic activity and excessively high circulating levels of catecholamines, demonstrates the author's long interest in and many studies on pheochromocytoma.

The book will be of interest to a wide variety of readers. For the clinician and cardiologist it tells clearly and precisely what is and what is not known about the subject and identifies the major areas of clinical interest. For the basic investigator it emphasizes many intriguing problems which remain to be solved and many important clinical applications of what is known about adrenergic physiology and pharmacology. It represents a very major effort and undoubtedly will make a most significant contribution to all who read it. Even though information will continue to accumulate and ideas to change, this work will surely stand the test of time.

Brian F. Hoffman, M.D.
David Hosack
Professor of Pharmacology and
Chairman of the Department of Pharmacology
College of Physicians and Surgeons of
Columbia University

I. Historical Background

In 1921, *Loewi* [243] made the important discovery that stimulation of the cardiac sympathetic nerves of the isolated perfused turtle heart resulted in an increase in the heart's rate and force of contraction. He further demonstrated that following stimulation of these nerves, the fluid perfusing the heart was capable of enhancing the rate and contraction of another isolated heart. His brilliant experiments firmly established that stimulation of cardiac sympathetic nerves resulted in the liberation of a chemical substance which could account for the chronotropic and inotropic effects on the heart. This then laid the foundation for the current concept of the mechanism of sympathetic nerve transmission.

About 12 years later, *Cannon and Rosenbluth* [52] demonstrated that the substance released by stimulating cardioaccelerator nerves entered the circulation and could cause a response in other organs (e.g. contraction of the nictitating membrane in the cat sensitized to cocaine). Since extracts of the heart produced biological effects which resembled those caused by epinephrine, *Cannon and Lissak* [51] concluded that sympathetic nerves contained epinephrine. A few years later *Simeone and Sarnoff* [358] found that contraction of the nictitating membrane following cardioaccelerator nerve stimulation could be prevented by administering the adrenergic blocking agent, dibenamine. During the same period *Hoffmann* et al. [183] observed that adding acetylcholine to an isolated mammalian heart preparation resulted in liberation of an epinephrine-like substance from the heart. However, it was *von Euler* [113] who, in 1946, established the fact that the sympathetic neurotransmitter (i.e. the sympathomimetic substance liberated during sympathetic nerve stimulation) was norepinephrine. Within the next decade *Goodall* [153] quantitated the content of norepinephrine and epinephrine in cattle heart and *Raab and Gigee* [319] reported studies on the concentrations of these biogenic amines in normal and diseased human hearts.

From an evolutionary viewpoint it is interesting that epinephrine and norepinephrine are present in high concentrations in chromaffin cells lining the heart cavities of cyclostomes and the Pacific hagfish, primitive vertebrates [29, 54]. As stated by *Braunwald* et al. [34] this finding provides strong morphologic evidence that even the hearts of organisms on a relatively low phylogenetic scale contain cells capable of secreting epinephrine and nor-epinephrine.

In the mammalian heart the neurotransmitter, norepinephrine, is located in storage vesicles in the sympathetic nerves and not in the myocardial cells. Observations by *Outschoorn and Vogt* [302] and *Siegel* et al. [357] revealed that stimulation of the cardio-accelerator nerves resulted in a marked elevation of the concentration of norepinephrine in coronary sinus blood and even some increase of the concentration in arterial blood. Then in 1962 and 1963 several investigators found that release of significant amounts of norepinephrine from the heart could be induced by a number of vasoactive agents (e.g. tyramine, quanethidine, reserpine, bretylium); further-more, the amount released was enough to cause sympathomimetic responses [60, 135, 136, 169].

Braunwald et al. [34] made additional observations which indicated that the norepinephrine released by electrical stimulation of the sympathetic nerves to the dog heart could elevate arterial blood pressure by a peripheral effect. They pointed out that: 'The overflow of norepinephrine into the venous blood is of greater magnitude in the coronary vascular bed than in other vascular beds for three reasons: First, there is a relatively large total quantity of norepinephrine in the heart. Secondly, the basal perfusion rate of the heart per unit weight of tissue exceeds that of the spleen, liver and limbs, organs which contain smaller although still substantial quantities of norepinephrine. This relatively high perfusion rate facilitates the delivery of norepinephrine released from sympathetic nerve endings into the coronary venous blood prior to its return to the nerve endings and before its enzymatic degradation. Finally, the vasodilatation in the coronary bed consequent to sympathetic nerve stimulation also aids passage of the neurohormone into the coronary venous blood, whereas the vasoconstrictor response to sympa-thetic stimulation in other vascular beds impedes the overflow of norepine-phrine into the venous blood.'

In light of the facts cited in the foregoing historical review plus evidence that the isolated mammalian heart is capable of biosynthesis and storage of norepinephrine [62, 370], *Braunwald* et al. [34] suggested the intriguing concept that the heart can function as a neuroendocrine organ.

During the past two decades there has been an explosion of information regarding catecholamine biosynthesis, storage, secretion, inactivation, receptors, and pharmacology. Considerable light has been shed on the important role catecholamines may play in a variety of diseases. In this monograph an attempt is made to present some of the current concepts of catecholamines and their effects on the mammalian heart. A potentially successful human cardiac transplantation (i.e. an extrinsically denervated heart which does not become reinervated) has caused some to deny the importance of neural regulation of the heart under normal physiological conditions. However, evidence has accumulated which now indicates that the autonomic nerves to the heart not only influence heart rate, contractile force, and atrial, ventricular and systemic blood pressures, but also velocities and patterns of conduction, electrophysiological properties of different portions of the heart, rhythmicity of contraction, and coronary blood flow. Furthermore, it seems likely that local metabolic reactions, membrane ionic mechanisms, and intracardiac reflexes may be modulated by either or both the sympathetic and parasympathetic cardiac nerves [322, p. 8].

II. Catecholamine Metabolism: Biosynthesis, Storage, Release, and Inactivation

A. General Remarks

1. Nomenclature, Occurrence, and Metabolism

Basic to an understanding of the function and pathophysiology of the catecholamines is a knowledge of the biosynthesis and inactivation of these biogenic amines. The term 'catecholamine' refers to any compound composed of a catechol nucleus (a benzene ring with two adjacent hydroxyl groups) and an amine-containing side chain; these substances are of low molecular weight. The catecholamines known to occur in man are dopamine, nor-epinephrine, and epinephrine. They are importantly involved in neural and endocrine function.

Dopamine appears to serve as a neurotransmitter in the central nervous system and to minor extent in some sympathetic ganglia. As will be mentioned later, there is accumulating evidence that dopaminergic nerves and receptors exert unique functions in the brain and elsewhere [147, 283, 284]; however, this amine functions as a precursor for norepinephrine. Epinephrine and norepinephrine are of major importance in affecting metabolism and cardio-vascular physiology. Biosynthesis of these amines occurs in the sympathetic neurons (mainly the nerve endings), brain, and chromaffin tissue. Both norepinephrine and epinephrine are synthesized in some chromaffin cells and parts of the brain, whereas only norepinephrine is synthesized in the post-ganglionic sympathetic nerves, where it serves the important function of the neurotransmitter (mediator of nerve activity) at most postganglionic sympathetic endings in the autonomic nervous system. All the enzymes necessary for conversion of the amino acid L-tyrosine to norepinephrine are present in the nerve ending, where the bulk of norepinephrine is synthesized and stored. Neurons which liberate catecholamines are called 'sympathetic' or 'adrenergic neurons.' In the fetus the adrenal [184, 416] and organ of Zuckerkandl (collections of chromaffin cells located anterior to the abdominal aorta just above its bifurcation and extending to the origin of the inferior mesenteric artery) [416, 417] contain only norepinephrine; however, epine-

Table I. Sites of norepinephrine storage in tissue

Organ	Cell	Relative amount	Present in granulated vesicle
Brain and spinal cord	adrenergic neuron:		
	cell body	moderate	no
	nerve ending	very large	partly
Sympathetic ganglia	adrenergic neuron:		
	cell body	small	no
Organs with sympa- thetic innervation (i.e. heart, spleen, liver, kidney, muscle, salivary gland)	adrenergic neuron: sympathetic nerve ending	very large (most of the norepinephrine in the body)	yes (adrenergic vesicle)
	extraneural pool (in parenchymal cells)	small	no
Adrenal medulla (and extramedullary chromaffin cells)	chromaffin cell	very large	yes (chromaffin granule)
Uterus	parenchyma (?)	moderate	no

The concentrations of norepinephrine in the adrenergic cell body and nerve ending have been estimated to be 10–100 and 10,000 μg/g, respectively. From *Wurtman* [428]; reprinted with permission.

phrine also appears in these organs within 1 year following birth [417]. The sites of norepinephrine storage in tissue are indicated in table I.

Although norepinephrine and epinephrine are secreted into the blood, where they can be demonstrated in minute concentrations in the free (un-conjugated) form, the presence of free dopamine in the blood of normal subjects was not quantitatively reported until 1973 [65]. More recently, with the use of a very sensitive assay, dopamine has been detected, although only sporadically, in a few normotensive and hypertensive subjects, and it has been suggested that the plasma concentration of this amine is independent of sympathetic activity [82]. In peripheral blood about 80% of norepinephrine and epinephrine and almost 100% of dopamine is conjugated [45, 224]. All three catecholamines are, however, present in urine in both the free and conjugated form. Therefore, it appears that dopamine in the urine results mainly from the conversion of dopa to dopamine in the kidney.

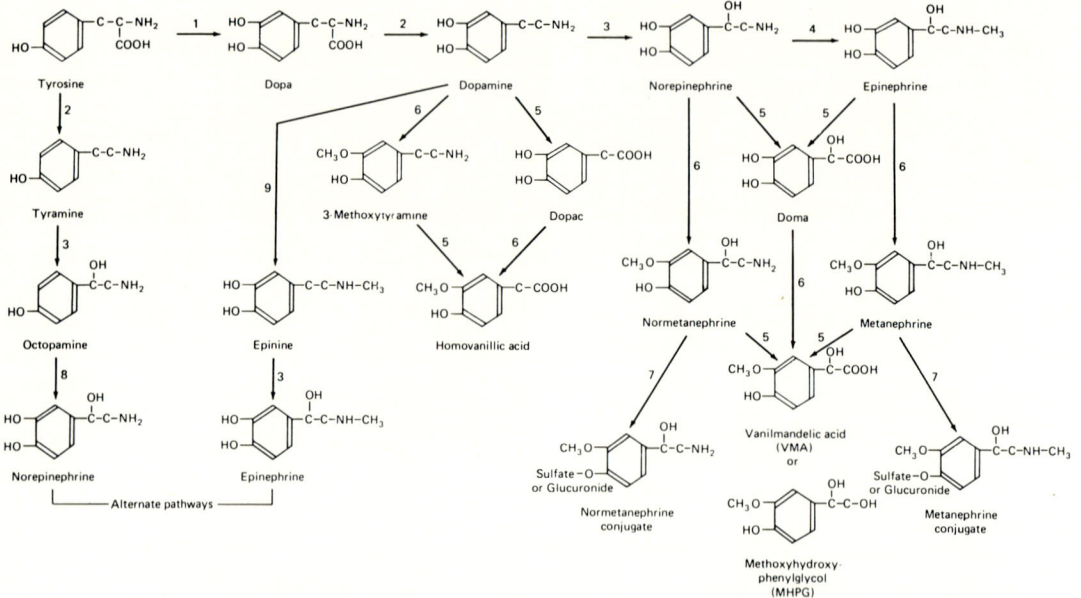

Fig. 1. Pathways of synthesis and metabolism of catecholamines with enzymes catalyzing various reactions. *1* Tyrosine hydroxylase; *2* aromatic amino acid decarboxylase; *3* phenylamine-β-hydroxylase; *4* phenylethanolamine-*N*-methyltransferase; *5* monoamine oxidase plus aldehyde dehydrogenase; *6* catechol-*O*-methyltransferase; *7* conjugating enzymes; *8* rabbit-liver enzyme; *9* rabbit-lung enzyme. From *Manger and Gifford* [255]; reprinted with permission.

Figure 1 reveals the pathways of biosynthesis and metabolism of catecholamines and indicates the enzymes catalyzing the various reactions. The biosynthesis, storage, release, and inactivation of the catecholamines in sympathetic nerves and chromaffin cells will now be considered.

B. Sympathetic Nerves

1. Biosynthesis, Storage, and Release of Norepinephrine

Figure 2 reveals a schematic representation of the biosynthesis, storage, and secretion of norepinephrine in a sympathetic nerve ending. Each sympathetic nerve may contain as many as 25,000 varicosities (or 'buttons'), which simply represent thickenings or bulges, at the nerve endings and along the course of the fibers, where norepinephrine is synthesized and stored in granulated vesicles. In addition to the norepinephrine contained in these

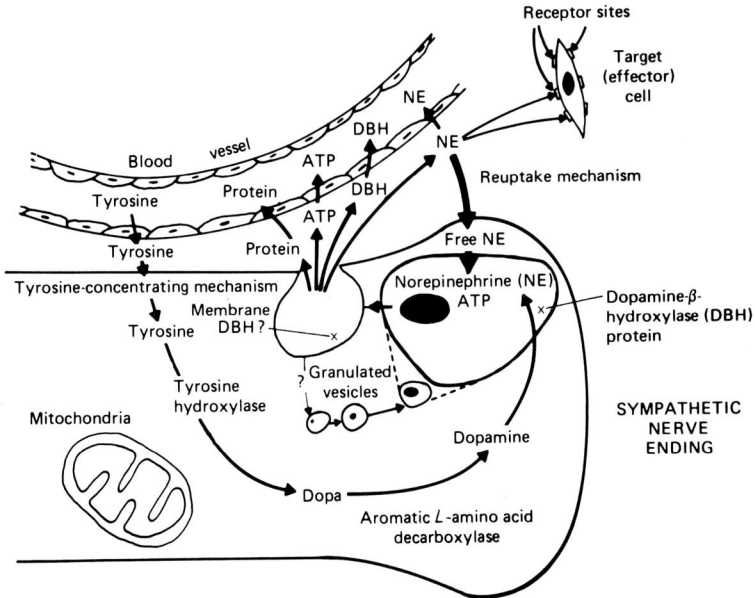

Fig. 2. Intracellular movements of substrates in the biosynthesis of norepinephrine; secretion via exocytosis of storage granules, and theoretical 'recycling' of granule. From *Manger and Gifford* [255]; reprinted with permission.

latter organelles (i.e. synaptic vesicles) an intracellular pool of 'free' norepinephrine is also assumed to exist [283, 284]. Most of the varicosities, which contain high concentrations of norepinephrine, lie in close proximity to effector (target) cells and represent synaptic regions of the sympathetic nerve terminals [177].

L-Tyrosine from the blood is thought to be transported across the membrane of the sympathetic nerves by a special concentrating mechanism. It is then converted by the enzyme tyrosine hydroxylase, which is found only within catecholamine-producing cells, to *L*-dihydroxyphenylalanine (dopa). This reaction, which requires tetrahydropteridine as a cofactor, proceeds slowly in vivo and is considered the rate-limiting step in the biosynthesis of the catecholamines [207, 236]. Tyrosine hydroxylase is inhibited by catecholamines and this inhibition appears important in controlling the rate of biosynthesis of norepinephrine in the sympathetic nerves [292]. Dopa, in turn, is rapidly decarboxylated to *L*-dihydroxyphenylethylamine (dopamine) by aromatic *L*-amino acid decarboxylase in the cytoplasm of the neurons. This second step proceeds rapidly and requires pyridoxal phosphate as a cofactor. Dopa decarboxylase is widely distributed even in tissues that do not normally synthesize catecholamines, and its high activity in kidney may explain the large amounts of dopamine found in urine.

Dopamine then enters minute granulated vesicles (400–600Å) in the sympathetic nerve terminals, where it is then finally hydroxylated by dopamine-β-hydroxylase (DβH) to *l*-norepinephrine. DβH is found only in cells that produce norepinephrine. This third enzymatic reaction requires O_2 and ascorbic and fumaric acids as cofactors. Norepinephrine, the neurotransmitter, remains inactive and protected in these storage vesicles until released by activation of the sympathetic nerves. A small amount of norepinephrine leaks from these granules into the surrounding cytoplasm, but these vesicles also have the capacity to take up and bind norepinephrine from the cytoplasm. It is believed that excitation of the sympathetic nerves results in a process of exocytosis (emiocytosis, or 'cell vomiting') whereby storage granules move, perhaps via microtubules, to the surface of the sympathetic nerve membrane, where they expel their contents into the extracellular fluid and circulation [414]. The contents of these vesicles consist of norepinephrine and adenosine triphosphate (ATP) in perhaps a 4:1 molar ratio, plus soluble DβH, and a small amount of protein (other than DβH) which is called chromogranin. It is possible that the empty vesicles are then reutilized (recycled) for the synthesis and storage of norepinephrine.

Primarily because of an efficient reuptake mechanism, a relatively small fraction of physiologically active neurotransmitter reaches receptors on the target (effector) cells and thereby activates these cells (e.g. vascular smooth muscle, myocardium, adipocyte, myometrium, or hepatocyte; receptors also exist at presynaptic neural sites). Activation of the sympathetic nervous system may sometimes cause no appreciable increase in the plasma concentration of norepinephrine [256]. This finding is probably explained by the efficient reuptake mechanism and enzymatic degradation which prevent norepinephrine from overflowing into the circulation in significant amounts. It is assumed that the relatively small amount of norepinephrine which reaches the circulation does so by a process of diffusion; however, there are no experimental studies which either validate or refute this concept.

The physiologic response of target cells to injected or secreted catecholamines depends on (1) the fraction of catecholamine delivered to the target cell (this can vary with the state of the circulation); (2) ability of the cell to inactivate the delivered catecholamines, and (3) sensitivity of the target cell [428, p. 33].

Release of norepinephrine and DβH is enhanced by calcium ions, and also by some alpha-adrenergic blocking agents (e.g. phenoxybenzamine and phentolamine). This enhanced release can be inhibited by prostaglandin E_2, perhaps due to prostaglandin's interference with availability of Ca^{2+} [16].

During the past few years evidence has accumulated for the existance of a presynaptic regulation of norepinephrine release from adrenergic nerves [229]. It has been postulated that norepinephrine released by nerve stimulation (once it reaches a threshold concentration in the synaptic gap) activates presynaptic alpha receptors, triggering a negative feedback mechanism which inhibits further release of the neurotransmitter. Compatible with this hypothesis is the fact that alpha receptor agonists inhibit, whereas alpha receptor antagonists enhance, norepinephrine release by nerve stimulation. It now appears that, in addition to alpha receptors in adrenergic nerve endings, there are dopaminergic and muscarinic receptors which are inhibitory to neurotransmitter release, and also nicotinic receptors which elicit norepinephrine release [229]. The existence of a presynaptic beta-adrenergic receptor seems to be an equally plausible concept: stimulation of a presynaptic beta receptor would enhance the norepinephrine release during adrenergic stimulation and thus constitute a positive feedback control of neurotransmitter secretion. Recently, *Stjärne and Brundin* [379] have demonstrated a dual adrenoceptor-mediated control of norepinephrine secretion from human vasoconstrictor nerves, i.e. a facilitation by beta receptors and an inhibition by alpha receptors in the presynaptic region. From their elegant experimental observations it was suggested that low sensitivity for alpha adrenoceptors can only be triggered by high concentrations of norepinephrine occurring in the synaptic cleft. On the other hand, the extremely high sensitivity of beta-adrenoceptors should enable them to detect physiologic concentrations of circulating catecholamines. These latter receptors may thus subserve the function of enhancing secretion of norepinephrine from synaptic nerves during conditions of increased secretion of epinephrine from the adrenal medulla. However, high epinephrine concentrations may depress norepinephrine secretion from synaptic nerves by stimulating the less sensitive alpha-adrenergic-mediated control mechanism.

It also appears that other presynaptic receptor systems exist which can modify catecholaminergic transmission: angiotensin II has been shown to increase the release of norepinephrine evoked by electrical stimulation of noradrenergic neurons, whereas some prostaglandins (PGE_1 and PGE_2) and some narcotic analgesics decrease release of the neurotransmitter [375].

For a detailed account of the local regulation of adrenergic neurotransmission one should consult the review by *Westfall* [418]. As pointed out by *Nadeau* et al. [291], most studies on this regulation of neurotransmission have been performed in tissue preparations; however, more recent experiments indicate that these regulatory mechanisms occur in the intact animal [230, 291, 431]. For example, alpha-adrenergic blockade with phenoxybenzamine prevents the negative feedback mechanism ordinarily exerted by norepinephrine on the presynaptic alpha receptor; consequently, during cardiac sympathetic nerve stimulation, there is a greater increase in catecholamine levels in coronary sinus blood in the animal with alpha-adrenergic blockade than in the control (fig. 3). Contrariwise, administration of clonidine, an alpha-adrenergic agonist, reduced the overflow of norepinephrine into coronary sinus blood caused by cardiac sympathetic nerve stimulation [291]. Administration of Sotalol (a beta-adrenergic blocker) reduced the overflow of norepinephrine into coronary sinus blood while isoproterenol (a beta-adrenergic agonist) increased norepinephrine overflow because of their respective effects on blockade and stimulation of the presynaptic beta receptor (fig. 4).

Infusion studies with epinephrine and norepinephrine suggest that the low concentrations of circulating catecholamines normally present have little or no effect in stimulating the intact heart [88]. However, high concentrations of circulating catecholamines occurring during strenuous exercise or cardiovascular shock or in persons harboring pheochromocytoma may cause very significant changes in cardiac function [255].

2. Inactivation of Norepinephrine

(a) *Catabolism.* Inactivation of norepinephrine occurs in several ways. The norepinephrine which is free in the cytoplasm of sympathetic nerves may be deaminated by monoamine oxidase (MAO) plus aldehyde dehydrogenase in mitochondria to form an unstable aldehyde, which can then be oxidized to an acid, 3,4-dihydroxymandelic acid (DHMA or DOMA); or norepinephrine may be reduced to an alcohol, dihydroxyphenylglycol (DHPG). The reduction product predominates in the rat, whereas the oxidation product predominates in man [18].

MAO, which is widely distributed and is particuarly abundant in the brain, liver, and kidney, was once considered to play a key role in terminating the physiologic action of the catecholamines. This enzyme is now thought to

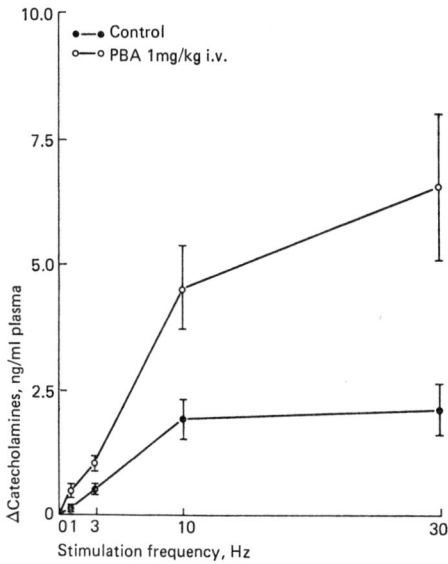

Fig. 3. Effect of phenoxybenzamine on coronary sinus blood noradrenaline levels.
From *Nadeau* et al. [291]; reprinted with permission.

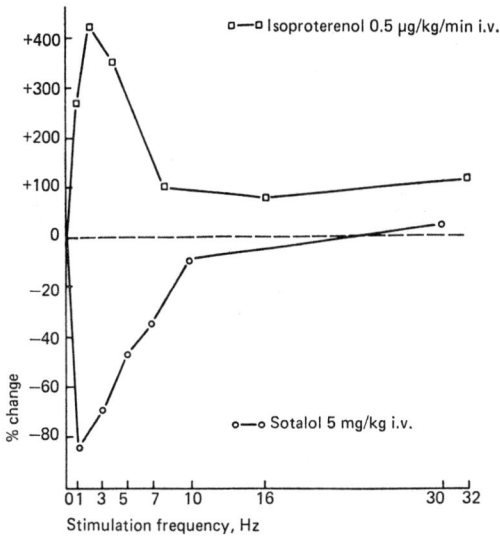

Fig. 4. Effect of sotalol and isoproterenol on coronary sinus noradrenaline levels.
From *Yamaguchi* et al. [431]; reprinted with permission of the American Heart Association.

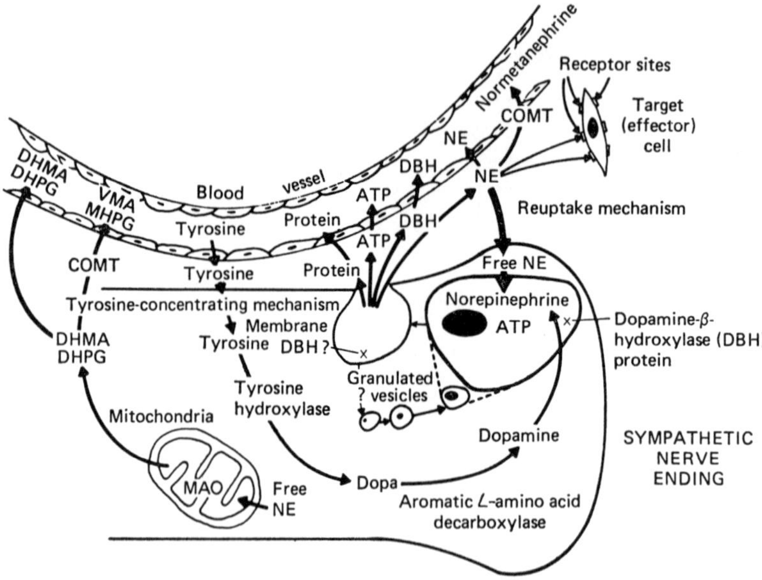

Fig. 5. Intracellular movements of substrates in the biosynthesis of norepinephrine; secretion via exocytosis of storage granules, and theoretical 'recycling' of granule, and catabolism. COMT = Catecholamine-*O*-methyltransferase (in cells); MAO = monoamine oxidase (in mitochrondria); DHMA = dihydroxymandelic acid; DHPG = dihydroxyphenylglycol; VMA = vanillylmandelic acid; MHPG = methoxyhydroxyphenylglycol. From *Manger and Gifford* [255]; reprinted with permission.

be primarily concerned with disposing of excess stores of catecholamines [219]. On the other hand, the norepinephrine which is released into the circulation is largely converted by *meta-O*-methylation to normetanephrine by the action of catechol-*O*-methyltransferase (COMT) [17, 226]. COMT, first identified by *Axelrod* [13], is present in almost all tissues and particularly concentrated in liver and kidney. Although COMT has been considered to be located in extraneuronal sites, there is recent evidence of an intraneuronal location in some tissues [145]; it does not, however, appear to be present in very significant amounts in sympathetic nerves. COMT requires *S*-adenosylmethionine as the methyl donor. The biosynthesis, storage, secretion, and catabolism of norepinephrine in the sympathetic nerve is diagrammatically illustrated in figure 5.

Further degradation of normetanephrine and metanephrine by MAO and of 3,4-dihydroxymandelic acid by COMT in cells elsewhere in the body results in the formation of 3-methoxy-4-hydroxymandelic acid (VMA). A small amount of methoxyhydroxyphenylglycol (MHPG) is generated by the enzymatic reaction of MAO with metanephrine and normetanephrine, and by the reaction of COMT with dihydroxyphenylglycol. Some free norepinephrine and normetanephrine are converted to glucuronide and/or sulfate conjugates [45] which are biologically inert, as are the metabolites formed by the action of MAO and COMT [117, 226].

(b) *Uptake and Binding.* The inactivation of catecholamines entering the blood is extremely rapid. For example, radioactive norepinephrine and epinephrine, when injected into animals in physiologic concentrations, have an initial half-life of 10–30 s [413, 419, 428]. *Vendsalu* [407] reported the half-life in plasma of norepinephrine infused into humans to be 2.3 min.

The fate of the catecholamines in the circulation depends largely on the organs to which they are delivered [428, p. 27]. Organs richly innervated with sympathetic nerves (e.g. heart and spleen) take up circulating catecholamines and bind most of them in a physiologically inactive form in the storage vesicles of sympathetic nerve endings. On the other hand, organs containing relatively high concentrations of COMT and MAO (e.g. liver and kidney) convert the catecholamines to their metabolites (i.e. normetanephrine, metanephrine, VMA, or MHPG). Because of the blood-brain barrier, only the hypothalamus takes up detectable amounts of catecholamines [413].

Manger et al. [257] and *LaBrosse and Hertting* [227] demonstrated the remarkable ability of the liver to clear 85% or more of the catecholamines from the circulation. The heart extracts 70 or 80% of these amines from the circulation during a single passage [429].

Rapid removal and physiologic disposition of catecholamines from the circulation cannot be entirely accounted for by sympathetic nerve uptake [399]. *Avakian and Gillespie* [11, 12] found that norepinephrine is taken up in vitro – not only by adrenergic nerves but also by smooth muscle, collagen, and elastic tissue – and they suggested that smooth muscle cells have a transport mechanism for intracellular norepinephrine uptake. *Kaumann* [208] obtained results consistent with a saturable extraneuronal mechanism with high affinity for norepinephrine. Even catecholamine entry into blood cells has been demonstrated [225, pp. 15, 16; 256a].

Thus, there are 'specific' (neuronal) and 'nonspecific' (extraneuronal) catecholamine storage sites. The uptake of catecholamines by sympathetic nerves (uptake$_1$) and extraneuronal uptake of catecholamines (uptake$_2$) have been reviewed and studied by *Iversen* [191]; however, the mechanisms and importance of extraneuronal catecholamine uptake are complex and have not been adequately investigated. The most important mechanisms for removing and inactivating circulating catecholamines are uptake into sympathetic nerves and metabolic degradation by the enzyme COMT [15].

(c) *Reuptake Mechanism.* Of major importance in terminating the physiologic action of norepinephrine released at the sympathetic nerve terminals is the neuronal reuptake process. This high-affinity neuronal uptake is stereoselective and requires sodium; it can be utilized by other amines structurally similar to norepinephrine (e.g. epinephrine, dopamine, tyramine, α-methylnorepinephrine, metaraminol, and amphetamine) [16]; however, the affinity for norepinephrine is about twice that for epinephrine and dopamine. These structurally related amines can also be taken up by the storage vesicles and displace norepinephrine. *Potter and Axelrod* [314] demonstrated that organs having extensive sympathetic nerve innervation take up and store circulating catecholamines in a chemically unchanged form in granulated vesicles of sympathetic nerve terminals. *Moskowitz and Wurtman* [283] have imaginatively likened the release and reuptake of the neurotransmitter to the behavior of a sponge: 'The wave of depolarization squeezes the metaphoric sponge, causing catecholamine molecules to drip out into synapses; with repolarization the sponge snaps back into shape, sopping up the catecholamines in the synaptic cleft.' More than 85% of the neurotransmitter in sympathetic nerves is synthesized locally, the rest being taken up from the circulating catecholamines [190].

Perhaps as much as 80% of norepinephrine liberated from synaptic nerves into the extracellular fluid is taken up into adjacent synaptic nerve terminals by an active transport mechanism [123]. Such a reuptake mechanism is of great value to the economy and efficient function of the sympathetic nervous system. To emphasize the importance of the reuptake process in terminating the action of norepinephrine, it should be appreciated that the biologic activity of the catecholamines is promptly terminated even if catabolic enzymatic degradation is pharmacologically blocked. On the other hand, blocking the reuptake mechanism by administering cocaine or desmethylimipramine will considerably prolong the physiologic response caused by sympathetic nerve stimulation [190]. Hypersensitivity of sympathetically denervated structures to catecholamines may partly be explained by degeneration of sympathetic nerves and disappearance of storage vesicles following nerve section. As a consequence, vesicles are not available to take up the catecholamines, and therefore an excess of these amines is available to stimulate the adrenergic receptors of the denervated structure.

C. Chromaffin Cells

1. Biosynthesis, Storage, Release, and Inactivation

The biosynthesis, release, and metabolism of norepinephrine in chromaffin cells is essentially the same as that in the sympathetic nerves; however, some chromaffin cells (e.g. certain cells in the adrenal medulla) have the capacity to convert l-norepinephrine to l-epinephrine. This reaction is brought about by the enzyme phenylethanolamine-N-methyltransferase (PNMT), which transfers a methyl group from S-adenosylmethionine to the nitrogen position of norepinephrine. In mammalian tissues, only the heart, adrenal medulla, brain [14] and brainstem [339] have measurable amounts of this epinephrine-synthesizing enzyme (PNMT). Since PNMT, when present, is located in the cytoplasm of chromaffin cells, norepinephrine must migrate from the chromaffin granule to the cytoplasm for methylation, and then return to the granule for storage.

It is noteworthy that enzymes (tyrosine hydroxylase, DBH, and PNMT) controlling catecholamine biosynthesis in the adrenal medulla are under hormonal control [16]. Furthermore, some recent evidence sugests that glucocorticoids modulate transsynaptic induction of tyrosine hydroxylase in sympathetic ganglia as well as in the adrenal medulla [301].

80–85% of the catecholamine content of the adult human adrenal medulla is epinephrine and 15–20% norepinephrine. Plasma and urinary levels of norepinephrine correlate reasonably well with sympathetic nerve activity, since norepinephrine enters the circulation mainly as an 'overflow' from the adrenergic nerves. Catecholamines released from the adrenal glands are inactivated mainly by COMT and MAO in the liver.

Bilateral adrenalectomy markedly reduces the concentration of epinephrine in the urine of man, whereas the norepinephrine concentration is not significantly altered [115, 150]. Any remaining epinephrine (perhaps one fifth of that normally excreted in the urine) is presumed to come from chromaffin cells present elsewhere in the body [114, p. 287]. It should be pointed out that the adrenal medulla can be viewed as a sympathetic ganglion which is innervated by preganglionic cholinergic fibers. These fibers release acetylcholine, which causes secretion of catecholamines by a process of exocytosis from the chromaffin cells of the adrenal medulla. The control of secretion from other chromaffin cells has not been adequately elucidated.

Secretion of catecholamines by a process of exocytosis was first studied in the adrenal medulla, where it appears to be responsible for the release of catecholamines from these chromaffin cells [94, 409]. It was further demonstrated that the presence of calcium was not only essential but was itself sufficient for the secretion of catecholamines in response to acetylcholine; other ions (potassium, sodium, chloride, and magnesium) were not necessary [94]. Secretion of adrenal catecholamines caused by other substances (e.g. histamine, serotonin, angiotensin and bradykinin) appears to depend on depolarization of the chromaffin cells and the entry of calcium, the latter being a critical event in stimulus-secretion coupling of chromaffin cells [94].

A similarly important role for calcium in stimulus-secretion coupling which involves exocytosis has been demonstrated in cholinergic and adrenergic nerves, where the respective neurotransmitters, acetylcholine and norepinephrine, are contained in membrane-limited structures (synaptic vesicles).

III. Adrenergic Receptors and Responses

Receptors can be viewed as target or binding sites on a cell which when activated by an agonist (i.e. a substance that interacts with a receptor and evokes a biological response) can cause a series of events that lead to a response in that cell. Receptor sites have never been seen and identified as anatomical structures; rather, their existence has been postulated on the basis of biological responses. Catecholamines are capable of activating receptor sites of a wide variety of cells. For example, catecholamines can cause lipolysis in adipose cells, glycogenolysis in liver and skeletal muscle, aggregation of platelets, chronotropic and inotropic effects on the myocardium, smooth muscle contraction or relaxation, glandular secretion, and even alter blood cell activity. The catecholamine receptors have been designated alpha-(α) and beta-(β) adrenergic receptors depending on the type of response caused by the catecholamines. Some typical adrenergic responses are listed in table II. Alpha and beta responses can also be distinguished by specific antagonists which block specific responses as schematically depicted in figure 6.

Beta-adrenergic receptors can be further subdivided into beta$_1$ (e.g. those which when stimulated cause a positive inotropic effect on the myocardium, lipolysis of fat cells, and inhibition of intestinal motility) and beta$_2$ receptors (e.g. those which when stimulated cause bronchodilatation, glycogenolysis, myometrial and smooth muscle relaxation). Epinephrine and norepinephrine are approximately equipotent in eliciting a beta$_1$ response whereas epinephrine is much more potent than norepinephrine in eliciting a beta$_2$ response. Certain antagonists (i.e. substances that interact with alpha or beta receptors and alter or occupy the receptors and thereby block biological responses) have a greater potency in blocking beta$_1$ than beta$_2$ receptors; however, there are no known antagonists which specifically block only beta$_1$ or beta$_2$ receptors [421]. Demonstration that the stereoisomers of epinephrine and norepinephrine with the levo (–) configuration are considerably more potent than those with the dextro (+) configuration indicated that a specific 3-dimensional configuration is required for stimulation of the receptor site to cause a maximal response. Beta- (–) adrenergic agonists are

Table II. Some typical adrenergic responses

Tissue	Alpha response	Beta response
Smooth muscle		
Uterus (rabbit)	contraction	relaxation
Pyloric sphincter	contraction	relaxation
Bronchial		relaxation
Bladder (detrusor)		relaxation
Bladder (trigone and sphincter)	contraction	
Iris (radial muscle)	contraction	
Ciliary muscle (lens)		relaxation
Intestine	decreased motility	decreased motility
Arterial	contraction	relaxation
Adipose tissue		lipolysis
Salivary glands	$K^+ + H_2O$ secretion	amylase secretion
Lymphocytes		inhibition of cytolysis
Cardiac muscle		
Contractility	increased	increased
Heart rate		increased
Functional refractory period	increased	decreased
Platelets	aggregation	inhibition of aggregation

From *Williams* and *Lefkowitz* [421]; reprinted with permission.

more stereospecific than alpha (–) agonists; beta (–) stereoisomers of the catecholamines are two to three orders of magnitude more potent than the corresponding (+) stereoisomers, whereas the alpha (–) stereoisomers are about one order of magnitude greater than their corresponding (+) stereoisomers [421]. Representative alpha- and beta-adrenergic compounds are listed in table III.

In addition to alpha- and beta-adrenergic receptors which respond to epinephrine and norepinephrine (table IV), a third type of adrenergic response caused by dopamine has been defined. Specific dopaminergic receptors have been identified in the brain and peripheral nervous system [79], and also in renal, mesenteric and cerebral vascular beds, where dopamine causes vasodilatation [146, 148, 149, 192]. Some evidence indicates that coronary vessels contain dopaminergic receptors which when stimulated can cause vasodilatation [348, 404]. *Goldberg* et al. [148, 149] have also shown that dopamine acts on beta$_1$ and alpha receptors and possibly on beta$_2$ and serotonin receptors.

The central nervous system and the peripheral nervous system contain several dopamine receptor subtypes; as mentioned previously, activation of

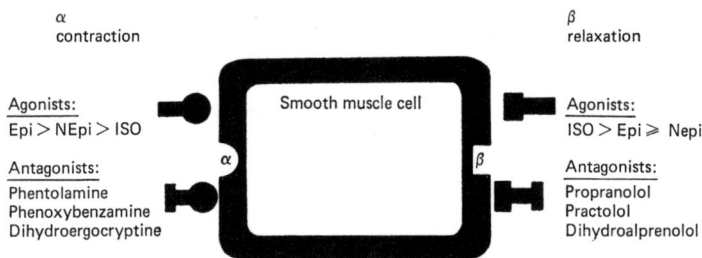

α
contraction

Agonists:
Epi > NEpi > ISO

Antagonists:
Phentolamine
Phenoxybenzamine
Dihydroergocryptine

Smooth muscle cell

α   β

β
relaxation

Agonists:
ISO > Epi ≥ Nepi

Antagonists:
Propranolol
Practolol
Dihydroalprenolol

Fig.6. The pharmacological differentiation of alpha- and beta-adrenergic responses. A schematic diagram of a typical smooth muscle cell is shown. (The response to alpha-receptor stimulation by an alpha agonist appears to result from a change in ion permeability whereas the response to beta-receptor stimulation by a beta agonist appears to involve accumulation of a cyclic nucleotide.) From *Williams and Lefkowitz* [421]; reprinted with permission.

Table III. Some representative alpha and beta adrenergic compounds

Alpha-adrenergic agonists	Alpha-adrenergic antagonists
Catecholamines	*Imidazolines*
Epinephrine, dopamine, norepinephrine, nordefrin, isoproterenol	phentolamine, tolazoline
Other phenylethylamines	*Haloalkylamines*
Phenylephrine, methoxamine, metaraminol, ephedrine, hydroxy- amphetamine, mephentermine	dibenamine, phenoxybenzamine
	Ergot Alkaloids
Ergot Alkaloids	ergotamine, dihydroergotamine, ergocryptine, dihydroergocryptine,
Ergonovine, ergotamine, methysergide	ergocrystine, dihydroergocrystine, ergocornine, dihydroergocornine
Others	*Others*
Clonidine	yohimbine, dibozane, phenothiazines, e.g. chlorpromazine, butyrophenones, e.g. haloperidol
Beta-adrenergic agonists	**Beta-adrenergic antagonists**
Hydroxybenzylisoproterenol, isoproterenol, epinephrine, norepinephrine, protokylol, cobefrin, salbutamol, soterenol, metaproterenol, isoetharine, phenylephrine	*Aryloxyethanolamines* alprenolol, propranolol, oxprenolol, practolol, hydroxybenzylpindolol, dihydroalprenolol
	Phenylethanolamines isoxuprine, ritodrine, nylidrin, butoxamine, dichlorisoproterenol

From *Williams* and *Lefkowitz* [421]; reprinted with permission.

Table IV. Effects of circulating norepinephrine (NE) and epinephrine (E) and receptors stimulated

Effector system	Response and type of receptor stimulated[a]	
	NE	E
Isolated heart	positive inotropic β and chronotropic β	positive inotropic β and chronotropic β
Heart frequency in vivo	bradycardia (vagal reflex)	tachycardia β
Mean arterial blood pressure	increase	slight increase or decrease
Skeletal muscle	vasoconstriction α	vasodilatation β
Liver	vasoconstriction α	vasodilatation β
Skin	vasoconstriction α	vasoconstriction α
Kidneys	vasoconstriction α	vasoconstriction α
Sweat glands (localized[b] secretion)	activation α	activation α
Intestinal smooth muscle (decrease of motility and tone)	relaxation α, β	relaxation α, β
Pupils	weak dilatation α	dilatation α
Central nervous system	slight or no effect?	apprehension, excitation?
Blood sugar (glycogenolysis)	slight increase β	increase β
Free fatty acids (lipolysis)	increase β	increase β
Basal metabolic rate (with increased heat production)	slight increase (mainly due to increased FFA)	increase (mainly due to increased FFA?)
Blood eosinophils	slight fall?	fall?

[a] The receptors on which NE and E act can be classified as α and β receptors depending on the reaction of the effector organs to contact with these amines. β-Receptors have been subdivided into B_1, and $β_2$ [132]. For example, cardiac inotrophy and lipolytic effects of the catecholamines are $β_1$ responses whereas bronchodilatation and vasodilatation appear to be $β_2$ responses. Evidence suggests that in distal coronary resistance vessels there are α- and β-adrenergic receptors which on stimulation may, under some circumstances, cause vasoconstriction and vasodilatation respectively [285].
[b] Stimulation accounts for activation of sweat glands on palms and a few other regions but generalized sweating in patients with pheochromocytoma remains unexplained. ?= Mechanism uncertain.
From *Manger and Gifford* [255], and taken partially from *von Euler and Ström* [116].

presynaptic dopamine receptors on postganglionic sympathetic nerves can inhibit norepinephrine release. Furthermore, dopamine inhibits prolactin release from the anterior pituitary gland, regulates beta-endorphin release from the intermediate pituitary, and can also cause release of parathyroid hormone [79]. Activation of dopaminergic pathways in the hypothalamus will elevate growth hormone levels [307] whereas dopamine is effective in lowering growth hormone levels in acromegalics by its direct action on the pituitary [408]. It is notable that an increase in dopamine receptor number appears central to Parkinson's disease, tartive dyskinesia, and possibly schizophrenia [79].

A high degree of binding specificity and a strong affinity for catecholamines characterize the receptor-catecholamine interaction. The interaction is rapid and reversible. Several thousand receptors have been demonstrated in a single cell. These receptors are macromolecules which appear to be proteins and which are located in the plasma membrane [232]. Stimulation of adrenergic receptors results in a sequence of biochemical events that produce a biologic response. In many instances the sequence appears to be the following: The catecholamine (a so-called 'first messenger') stimulates the enzyme adenylate cyclase which is located on the internal surface of the plasma membrane of the target cell. This enzyme then accelerates an intracellular generation of a cyclic nucleotide (the so-called 'second messenger') which in turn activates enzymes known as protein kinases; the kinases then phosphorylate a wide variety of important substrates which appear to mediate the characteristic responses attributed to the catecholamines.

Beta-adrenergic responses are usually accompanied by an increase in the formation of cyclic AMP which is thought to mediate the cellular response. In contrast, alpha-adrenergic stimulation is usually associated with a decrease or no change in cyclic AMP (with the exception of the brain, where increases have been reported in some regions). Therefore, the responses to alpha- and beta-receptor stimulation appear to be evoked by different biochemical mechanisms [421].

The number of functional receptors can be influenced by hormonal, genetic, and developmental factors. Changes in the number of receptors may profoundly alter sensitivity and responsiveness to catecholamines [421]. Prolonged exposure of receptors to their agonists attenuates the biologic response to that agonist. This desensitization (which actually is never complete) has been demonstrated with a variety of hormones and drugs. It is reversible, since resensitization occurs when receptor exposure to an agonist is terminated. Studies with beta-adrenergic receptors revealed that desensitization was accompanied by a reduction in a number of receptor sites and a decrease in responsiveness of adenyl cyclase to the agonist. Although this apparent loss of receptor sites may be explained by an increased avidity of the agonist for binding which thereby blocks further physiological response [363], there is little reason at the present time to believe that this is necessarily a general principle [*Lefkowitz*, personal communication]. It is noteworthy that antagonists, such as propranolol, do not desensitize beta-adrenergic receptors.

The precise mechanism responsible for desensitization remains unclear; however, the process of desensitization of cells exposed to beta-adrenergic agonists appears to be an important means of suppressing cellular responsiveness in the face of chronic and excess exposure to beta-adrenergic agonists. As pointed out by *Lefkowitz* et al. [233], the ability for hormone-sensitive cells to regulate their sensitivity and responsiveness may be important in the maintenance of cellular homeostasis – a regulation that 'permits adjustment of responsivity to a level appropriate for the ambient hormonal environment'. They cited the evidence for subsensitization (desensitization) for beta-adrenergic stimulation of adenyl cyclase in adipocytes and lymphocytes in patients with excess circulating catecholamines resulting from a pheochromocytoma or therapy with beta-adrenergic agonist drugs [233].

Although much less is known about alpha-adrenergic receptors, desensitization to epinephrine has clearly been demonstrated in one model system where membrane voltage appeared to be critically involved [421].

As stated by *Snyder* [363], manipulations which could reverse desensitization would have important clinical implications. It is interesting that guanine nucleotides (guanosine triphosphate and guanosine diphosphate) can reverse the decreased beta-adrenergic binding that follows prolonged exposure to catecholamines. These nucleotides appear to selectively reverse the formation of a high affinity receptor complex of agonists for beta-adrenergic receptors, as well as receptors for glucagon, prostaglandin, dopamine, opiates, and alpha-adrenergic agonists. Other nucleotides (e.g. GMP, ATP, ADP or AMP) do not alter receptor-agonist affinity. *Snyder* [363] points out that: 'Alpha-adrenergic receptors and opiate receptors are regulated by guanine nucleotides in the same way as beta-adrenergic and glucagon receptors, but they are linked to decreases rather than increases in cyclic AMP.' It should be emphasized that the phenomenon whereby guanine nucleotides resensitize desensitized receptors has only been demonstrated in isolated membranes and does not appear to be a general mechanism in terms of reversing desensitization which is induced in intact cells [*Lefkowitz*, personal communication].

In contrast to the decreased sensitivity resulting from receptor exposure to agonists, supersensitivity of beta-adrenergic receptors occurs when blood or tissue catecholamines are significantly lowered. This supersensitivity was related to an increased number of beta-adrenergic receptors without any increase in affinity for the receptor site. Depletion of catecholamines in the rat heart resulted in a 50–100% increase in beta-adrenergic receptor number, and this was associated with increased generation of cyclic AMP in response to catecholamine stimulation in the perfused rat heart [421]. It is noteworthy that chronic treatment of rats with the beta-adrenergic antagonist, propranolol, caused a 100% increase in the number of beta-adrenergic receptors [143]. The intriguing question has been raised by *Williams and Lefkowitz* [421] as to whether an increased number of beta-adrenergic receptors due to propranolol treatment might account for the 'propranolol withdrawal syndrome' observed clinically after cessation of treatment.

Hormones can significantly influence the receptor number and responsiveness to catecholamines in a number of different target cells. For example, adrenalectomy in the rat is accompanied by a 3- to 5-fold increase in beta-adrenergic receptor number and an increased responsiveness to catecholamines; these changes are reversed by administration of cortisone. Steroids can also influence myometrial function. Catecholamines induce contraction of the myometrium via an alpha-adrenergic mechanism whereas inhibition of contraction or relaxation of the myometrium occurs via a beta-adrenergic mechanism. When the uterus is under the influence of estrogen, the alpha mechanism predominates and catecholamines cause contraction; however, administration of progesterone significantly inhibits the contraction effect caused by alpha-adrenergic stimulation. It seems probable that alteration of the alpha-receptor number accounts for these results. Alpha-receptor number is markedly reduced in the myometrium under the influence of progesterone when compared with the uterus predominantly under the influence of estrogen. No change in beta receptors is induced by either of these hormones nor is there any change in the affinity of catecholamines for receptor sites [421].

With regard to the heart, it is noteworthy that triiodothyronine can increase the sensitivity of the fetal mouse heart to the chronotropic effect of beta-adrenergic stimulation. Furthermore, a marked hyperresponsiveness of rat myocardial phosphorylase to catecholamines in hyperthyroidism has been observed. Administration of triiodothyronine or thyroxine to induce hyperthyroidism resulted in a very significant increase in the number of beta receptors without a change in receptor affinity for agonists or antagonists. *Williams and Lefkowitz* [421] have found no change in receptor sites in adipose cells from hyperthyroid rats; therefore, it is not clear whether receptor alteration occurs in tissues other than the myocardium in hyperthyroidism. The mechanism whereby hormones alter receptors remains unclear.

A. Adrenergic Receptors of the Myocardium

Catecholamine receptors in the heart consist of those in the myocardium and those in the coronary vessels.

The dominant adrenergic receptors in the heart appear to be of the $beta_1$ type and are located on the cardiac sarcolemma. These receptors are recognized and activated by catecholamines to varying degrees, depending on the affinity and binding at the receptor site. Activation of these myocardial receptors augments myocardial contraction and heart rate as well as the demand for oxygen in the heart. However, in addition to beta receptors,the sinoatrial node appears to possess alpha receptors which when stimulated exert a negative chronotropic response [195]. Some investigators have also proposed the existence of atrial myocardial alpha receptors which may subserve a negative inotropic function [187]. Others have suggested that the atria contain both alpha and beta receptors which respectively respond to stimulation by increasing or decreasing action potential duration [159].

Adrenergic agonists and glucagon stimulate the enzyme adenyl cyclase in the sarcolemma and thereby increase generation of cyclic AMP (the so-called 'second messenger'). The enhanced myocardial contractility results from an increased calcium influx across the sarcolemma. In addition, there is an accompanying acceleration of relaxation which appears to be brought about by activation of the calcium pump in the sarcoplasmic reticulum [206]. Accelerated relaxation (i.e. an abbreviated systole) is, of course, an essential requirement to permit adequate diastolic filling and an efficient cardiac output during conditions of enhanced myocardial contractility and tachycardia [206].

Figure 7 schematically depicts the interrelationships between calcium and cyclic AMP within the myocardium. The beta-adrenergic agonists stimulate the generat on of 3'5'-cyclic AMP which is then converted by the enzyme phosphodiesterase to 5'-AMP (a biologically inactive substance) and inorganic phosphate. It is noteworthy that phosphodiesterase inhibitors can exert catecholamine-like effects on the heart. The level of cyclic AMP in the cell is thus controlled by the rate of its synthesis and degradation. An increased intracellular concentration of cyclic AMP is associated with an enhanced influx of calcium ions into the cell. However, an increased concentration of intracellular calcium both inhibits adenyl cyclase and activates phosphodiesterase. Hence, the enhanced flow of calcium ions into the cell caused by cyclic AMP can ultimately lead to a decline in the level of this cyclic nucleotide by a negative feedback mechanism [206].

Recently it has become apparent that before a cellular response can be

β-adrenergic agonists

$\oplus$

$$\text{ATP} \xrightarrow[\text{cyclase}]{\text{adenylate}} \text{Cyclic AMP} \xrightarrow[\text{diesterase}]{\text{phospho-}} \text{AMP}$$

$\ominus$ $\oplus$ $\oplus$

Ca^{2+}

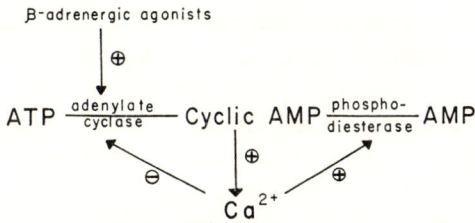

Fig. 7. Interrelationships between Ca^{2+} and cyclic AMP within the myocardium. Beta-adrenergic agonists activate adenylate cyclase, thereby causing cyclic AMP levels to rise. This in turn increases intracellular Ca^{2+}, which tends to reverse these effects by promoting a reduction in cyclic AMP levels through the ability of Ca^{2+} to inhibit adenylate cyclase and to activate phosphodiesterase. From *Tada* et al. [383]; reprinted with permission of the American Heart Association.

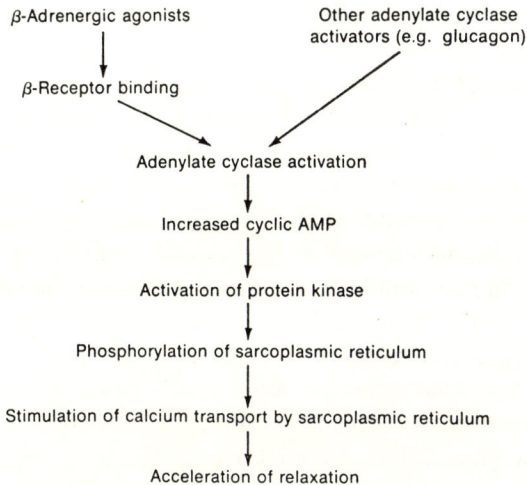

β-Adrenergic agonists   Other adenylate cyclase
                                    activators (e.g. glucagon)

β-Receptor binding

Adenylate cyclase activation

Increased cyclic AMP

Activation of protein kinase

Phosphorylation of sarcoplasmic reticulum

Stimulation of calcium transport by sarcoplasmic reticulum

Acceleration of relaxation

Fig. 8. Cascade of reactions by which agents that increase cyclic AMP levels can accelerate relaxation in the heart. From *Katz* [206]; reprinted with permission.

achieved, cyclic AMP must react with additional enzymes (protein kinases) to form still another messenger, the cyclic AMP-dependent protein kinases. Activation of the enzyme protein kinase then results in phosphorylation of the sarcoplasmic reticulum, thereby causing stimulation of calcium transport and an accelerated relaxation of the myocardium. This cascade of reactions is schematically represented in figure 8.

Dephosphorylation, which appears to limit and terminate the physiological effects caused by the cascade of reactions, is attributed to hydrolysis induced by a class of enzymes known as phosphoprotein phosphatases. Although cyclic AMP seems to be involved in the electrophysiologic responses of the heart to catecholamines, evidence for membrane dephosphorylation is inconclusive [206].

The precise mechanism of the excitation-contraction coupling is complex and incompletely understood. However, it appears that depolarization of the sarcolemma by an action potential initiates an influx of calcium into the interior of the myocardial cell. Calcium finally binds to troponin c, the calcium receptor of cardiac contractile protein, and contraction occurs [206]. The effects of norepinephrine which can be released from the sympathetic nerve terminals in all parts of the heart and of epinephrine and norepinephrine released into the circulation from the adrenal gland or elsewhere are similar. The role of the catecholamines becomes particularly important under conditions of physiologic and pathologic stress and under some circumstances of myocardial damage and congestive heart failure.

Braunwald et al. [36, p. 279] stated that: 'The quantity of norepinephrine released by sympathetic nerve endings in the heart is probably the most important mechanism regulating the position of the force-velocity and ventricular performance curves under physiological conditions.' They pointed out that almost instantaneous changes in myocardial contractility are affected by variations in impulse traffic in cardiac adrenergic nerves. Drugs which block these nerves or deplete catecholamine stores in the heart can block or reduce the myocardial response to stimulation of these nerves. Conversely, postganglionic denervation or drugs which block or impair neuronal reuptake of norepinephrine can increase the amount of norepinephrine at myocardial adrenergic receptor sites and thereby augment the response of the heart.

An infusion of isoproterenol, a synthetic catecholamine, stimulates the beta-adrenergic receptors of the heart in a manner quite similar to that of exercise (i.e. a reduction in afterload, in end-diastolic and end-systolic dimensions, accompanied by an increase in myocardial contractility, cardiac index, heart rate, and ventricular force-velocity relation). During exercise, beta blockade with propranolol 'reduced the endurance for maximal activity, the cardiac output, the mean arterial pressure, the left ventricular minute work, and the maximal oxygen uptake and increased the arteriovenous oxygen difference and the central venous pressure in normal human subjects' [36]. During exercise ventricular end-diastolic dimension did not decrease

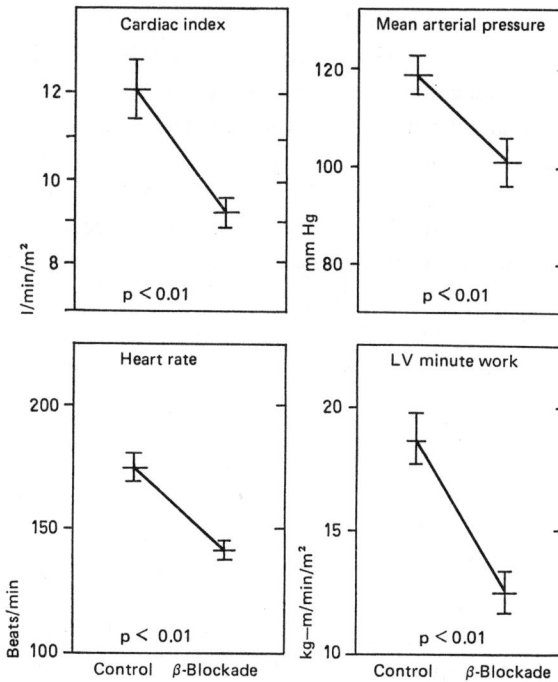

Fig. 9. Effects of beta-adrenergic blockade on four circulatory variables during maximal exercise in normal subjects. The mean values (±SEM) are shown for each variable. From *Epstein* et al. [108]; reprinted with permission.

following beta blockade, in contrast to the decrease noted in the unblocked state. The effects of beta-adrenergic blockade on four circulatory variables during maximal exercise is indicated in figure 9.

Various forms of stress can cause liberation of catecholamines into the circulation from the adrenal medulla and other portions of the adrenergic system. The response of the myocardium to these circulating catecholamines is less rapid in onset than the response following cardiac adrenergic nerve stimulation; however, it is identical to that caused by the neurotransmitter released from the adrenergic nerves in the heart.

B. Adrenergic Receptors of the Coronary Circulation

Regulation of the coronary circulation and the coronary artery response to catecholamines is difficult to study and interpret. Administration of catecholamines can induce: (1) an increase in systemic blood pressure and coronary artery perfusion pressure which may be accompanied by myogenic autoregulation in the coronary arteries which elevates coronary resistance; (2) an augmented myocardial contractility, due to beta$_1$ receptor stimulation, which increases myocardial oxygen consumption and metabolism and release of metabolites which cause coronary vasodilatation; (3) an increase or decrease in extravascular compression of the intramyocardial coronary vessels, depending on whether the heart rate is increased or decreased; (4) stimulation of alpha- and beta$_2$-adrenergic receptors which cause vaso-constriction and vasodilatation, respectively. As pointed out by *Braunwald* et al. [36, pp. 223–224], it is difficult if not impossible to predict the net effect of these various actions of the catecholamines on the coronary circulation and its blood flow.

Both alpha- and beta-adrenergic receptors have been identified in coronary arteries. Although most studies suggest that the beta-adrenergic receptor is beta$_2$ [36], some evidence suggests it could be a beta$_1$ receptor [22]. Furthermore, it is possible that dopaminergic receptors also exist in the coronary arteries since, in the presence of both alpha and beta blockade, dopamine causes coronary vasodilatation [404].

Identification of receptors can be facilitated by using isolated segments of coronary arteries and thus avoiding the influence on the coronary circu-lation of alterations in myocardial contractility and metabolism. Stimulation of alpha receptors of an isolated coronary segment causes vasoconstriction whereas beta$_2$-receptor stimulation induces vasodilation [22, 440]. The neurotransmitter, norepinephrine, which stimulates both alpha and beta receptors, causes relaxation in small coronary arteries; beta blockade will prevent this relaxation and permit only vasoconstriction due to alpha-receptor stimulation. In larger coronary segments norepinephrine usually causes an initial vasoconstriction followed by vasodilatation. The vaso-constriction can be prevented with an alpha-receptor blocker such as phenoxy-benzamine [440].

Stimulation of the stellate ganglion in the dog, which causes release of norepinephrine from cardiac sympathetic nerves, or intracardiac admini-stration of norepinephrine, causes alpha-receptor stimulation and a short period of vasoconstriction followed by vasodilatation, which results from the increased metabolic activity of the heart [25, 26, 36]. Following beta$_1$-adrenergic blockade (which prevents increments in heart rate, in myocardial contraction and in myocardial metabolism) stimulation of the sympathetic nerves to the heart produces only alpha stimulation and coronary vaso-constriction. The latter effect can be inhibited by alpha-adrenergic blockade

[118]. Some evidence suggests that the coronary arteries are normally under a tonic constriction which is mediated by the sympathetic nerves [402].

In summary, it appears that the coronary vasodilatation which follows administration of norepinephrine or sympathetic nerve stimulation is caused by the increased myocardial metabolism which accompanies beta$_1$-receptor stimulation of the myocardium. In the absence of increased myocardial metabolism, the alpha-constrictor response of the coronary arteries to the neuronally released or injected norepinephrine appears to be dominant [36]. *Braunwald* et al. [36] have pointed out that the coronary vascular responses to catecholamines and drugs may be markedly different in the unanesthetized as compared with the responses in anesthetized animals. Undoubtedly many conflicting results reporting the effect of catecholamines and drugs on the coronary circulation may depend on whether or not the animal was anesthetized.

IV. Neural Regulation of the Heart[1]

A. Innervation

Randall and Armour [323] have emphasized the importance of correlating function with neural structure innervating the heart at the time when the structure is identified. By so doing, errors in identifying cardiac nerves and in interpreting their function will be minimized. However, this approach of study is only possible in the experimental animal. For that reason we have described below the effects which *Randall* and his coauthors [322] observed when various cardiac nerves in the dog were electrically stimulated. For a detailed description of the nervous connections to the heart, the reader is referred to chapters 2, 3, and 4 by *Randall and Armour* [323] and *Levy* [237].

1. Cardiac Nerves on the Right in the Dog

In the dog two vagi (parasympathetic) nerves descend from the nodose ganglia through the neck where they become intermingled with sympathetic nerve fibers connecting the superior cervical and middle (caudal) cervical ganglia. Sympathetic preganglionic nerves leave the spinal cord from the second, third and fourth (and occasionally the first and fifth) anterior thoracic roots and join the stellate and middle cervical ganglia before going to the heart. Fibers from the middle cervical ganglia extend caudally and posteriorly to connect with the stellate ganglia. The latter connect with the sympathetic chains which extend caudally. The distribution of upper thoracic anterior roots to the dog myocardium and their contribution to inotropic responses are indicated in figure 10.

The *right stellate cardiac nerve* arises from the stellate ganglion (or its branches) and innervates the right atrium and provides a major innervation of the sinoatrial nodal region. Stimulation of the right stellate nerve can markedly augment sinus node pacemaker activity and induce near maximal obtainable heart rates. It can also augment inotropism in the atria, but it exerts little if any effect on the ventricles.

[1] Except for the section on 'Cardiac Denervation and Reinnervation', the author has relied almost totally on the views and concepts expressed by various contributors to 'Neural Regulation of the Heart' (Edited by *Walter C. Randall* [322]).

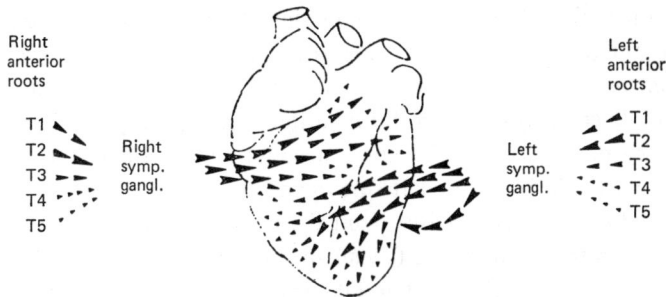

Fig. 10. Distribution of upper thoracic anterior roots to the dog myocardium. The size of the arrow heads indicate the relative contribution of anterior roots to regional inotropic cardiac responses. The right anterior roots dominate in the basal regions of the heart while the left roots dominate in the apical regions. From *Wurster* [426]; reprinted with permission.

The *right recurrent cardiac nerve* arises from the recurrent laryngeal nerve, the middle cervical ganglion and the vagus. It contains vagal and sympathetic efferent fibers and many afferents which arise from all four heart chambers; it has a mixed efferent and afferent function.

Two major cardiac branches, the *craniovagal and caudovagal nerves*, arise from the right thoracic vagus and innervate the right atrium. Stimulation of these nerves causes mainly a parasympathetic response, although both vagal and sympathetic effects (especially on the right side of the heart) may occur.

Randall and Armour [323] found that stimulating various *cardiac branches of the vagus nerve* on the right caused a wide variety of results. For example, stimulation at one level below the middle cervical ganglion caused a greater negative inotropism in the right than the left atrium and a suppression of conduction through the AV node but no atrial slowing (i.e. no effect on the SA node). On the other hand, with stimulation of an adjacent level atrial suppression was more prompt and more profound in both atria, and was accompanied by complete AV blockade; sometimes complete inhibition of SA nodal discharge was induced and asystole occurred. At other times atrial fibrillation occurred during or following stimulation. They concluded that different responses were due to different contents of parasympathetic and sympathetic nerve fibers innervating localized portions of the heart.

2. Cardiac Nerves on the Left in the Dog

The *innominate nerve* arises mainly from the middle cervical ganglion; it also receives a branch from the vagus and it may connect with the adjacent ansae-ganglion region. In addition, it contains afferent fibers from cardiac receptors. Stimulation induces both sympathetic and parasympathetic effects.

The *ventromedial nerve* arises from the vagus but receives a connection from the middle cervical ganglion. Stimulation causes parasympathetic and sympathetic cardiac effects – the latter accounting for enhanced cardiac contractility, particularly on the left side of the heart.

The *dorsal cardiac nerve* arises from the middle cervical ganglion and may connect with the vagus and a few small sympathetic branches. It sends only minor connections to the heart and it terminates in a fanlike distribution to the aortic arch and descending aorta. Although stimulation of this nerve may augment myocardial contractility, it appears that it contains numerous afferent fibers which reveal bursts of action potentials synchronous with the aortic pressure pulse.

The *ventrolateral cardiac nerve* arises from the middle cervical ganglion and sometimes partly from the adjacent ansa. It also has a few connections with the vagus and it receives a few branches from the recurrent laryngeal nerve. Its branches penetrate the pericardium to splay out upon the left atrium and a major portion extends over the ventral and dorsolateral left ventricular surfaces. It also forms a dorsal ventricular plexus near the coronary sinus. It has few afferent fibers and is considered primarily a sympathetic nerve which affects a large portion of the ventricular mass and the AV nodal region.

The *left stellate cardiac nerve* arises from the ansa and/or stellate ganglion and may become incorporated in the ventrolateral cardiac nerve. The nerve appears to contain primarily afferent fibers from the left atrium but occasionally a few from the ventricles; however, rarely, it may augment force of contraction or heart rate in the left atrium.

In addition to the large cardiac nerve fibers on the left, described above, small branches of the left vagus connect with the heart. Stimulation of these elicit profound parasympathetic effects, e.g. negative chronotropic effects on the heart and/or dromotropic changes involving the SA and/or AV nodal regions. Some afferents are also present in these nerves. As on the right side, no functional efferent cardiac nerves were found arising from the sympathetic chain below the stellate ganglion.

3. Peripheral Distribution of Cardiac Nerves

Although the myocardium is not anatomically a syncytium, the concept of a physiologic syncytium remains since excitation spreads throughout the entire heart [322, p. 78]. However, it has been demonstrated that stimulation of individual thoracic cardiac nerves results in contraction of localized myocardial segments. It appears, therefore, that specific nerves innervate specific muscle segments [382]. The effects of nerve stimulation on patterns of contractile force and changes in refractory periods in various parts of the myocardium have permitted identification and mapping of regional cardiac nerve distribution [322, pp. 78–85; 324].

4. Intracardiac Nerves

There are some neural elements present between myocardial cells which persist after extrinsic cardiac denervation. These elements appear to be primarily postganglionic parasympathetic elements although some may be sympathetic nerves which are associated with intrinsic ganglion cells. The atria, the SA and AV nodal regions, and the bundle of His receive a rich supply of excitatory adrenergic fibers and inhibitory cholinergic fibers. It is established that the SA node is supplied by sympathetic and parasympathetic nerves on the right side whereas the AV node is innervated by sympathetic nerves and the vagus on the left. The SA node contains perhaps the richest network of fibers in the heart [7], and there is evidence suggesting that each myocardial fiber in this region receives a separate innervation [80]. In addition, very high concentrations of catecholamines [356] and cholinesterase [198] are present in the SA node.

Elsewhere in the heart it appears that one cholinergic nerve terminal supplies a single effector muscle cell whereas one adrenergic nerve fiber releases neurotransmitter for several muscle cells. Nerves with vesicles containing granules have been considered adrenergic because they release norepinephrine. Agranular vesicles are characteristic of cholinergic fibers which release acetylcholine. However, unequivocal differentiation of cardiac afferent and efferent nerve types is difficult on a histologic basis. Some controversy still exists as to whether nerves terminate directly on myocardial cells. Although most investigators believe there is a synaptic cleft and that the neurotransmitter must diffuse from the nerve terminal to the myocardial effector cells, some studies have demonstrated neuromuscular contacts in some parts of the heart (e.g. AV nodal tissue). Detailed reviews of the fine structure of cardiac innervation have been presented elsewhere [30, 432].

The ventricles are richly innervated by adrenergic nerves (more so in the basal portions than in the apex) but the density of these nerves is much less than in the atria. Cholinergic innervation is sparse to moderate except in the proximal ventricular conducting system. *Kent* et al. [212] concluded from their studies that the rich cholinergic innervation of the ventricular conduction system in canine and human hearts may protect against spontaneous ventricular fibrillation during myocardial infarction. They also found that the increased ventricular fibrillation threshold caused by vagal stimulation was independent of adrenergic innervation. The anatomic arrangement of cardiac nerves in man is quite similar to that of the dog except that in man the sympathetic and vagal nerves are anatomically separate. However, a major connection between the vagus and middle cervical ganglia may occur [322, p. 28].

Sympathetic innervation of the dog heart is incomplete at birth, and the endogenous norepinephrine content of the heart does not reach adult levels until about the 56th day of life; however, the ability of the sympathetic innervation to take up norepinephrine occurs earlier and a normal (adult) positive chronotropic response to norepinephrine administration occurs by the 10th day of life [138].

B. Sympathetic Regulation

As pointed out by *Randall* [322, p. 45], autonomic innervation of the heart permits rapid and highly specialized adjustments in cardiac action. He stated that: 'There is compelling evidence that these neural mechanisms exercise greater control than hormonal (adrenal medulla) or intrinsic (heterometric) length-tension relationships.'

Exercise is accompanied by activation of the sympathetic nerves, augmentation of myocardial contractile force, and an elevated heart rate. An increased rate of myocardial tension development, a faster ejection velocity, and a shorter systole occur without a reduction in stroke volume. Ventricular contraction is much more rapid and the rates of change in pressure (dp/dt) are much greater than at rest. Augmentation of ventricular contraction causes an elevation of pressure pulse.

More intense exercise can invoke the Frank-Starling mechanism [322, p. 48]. *Siegel* et al. [357] reported elevations in the concentrations of catecholamines in coronary sinus blood which was related to increased activation of the sympathetic nerves innervating the myocardium. Depletion of cardiac catecholamines (which can occur in congestive heart failure or can be induced with a drug such as reserpine) results in a decreased inotropic response to sympathetic nerve stimulation; improvement of ventricular performance, which normally characterizes exercise, does not occur or is attenuated [322, p. 48].

Table V. The mean values of hemodynamic parameters and plasma catecholamine levels at rest and during isometric handgrip exercise (IHG)

Parameters	Number of subjects	At rest	During IHG
Age, year	22	41 ± 5.4	
Heart rate, beats/min	22	81 ± 5.4	98 ± 6.0
Mean aortic blood pressure, mm Hg	22	104 ± 4.2	130 ± 7.6
Cardiac output, L/min	18	6.5 ± 0.64	7.0 ± 0.69
Coronary sinus			
Blood flow, ml/min	8	149 ± 16	216 ± 35
Plasma flow, ml/min		88 ± 8.7	122 ± 12.5
Plasma catecholamines, pg/ml			
Coronary sinus	17		
Total catecholamine		485 ± 52	872 ± 208
Norepinephrine		359 ± 49	651 ± 91
Epinephrine		126 ± 18	221 ± 65
Aorta	22		
Total catecholamine		444 ± 47	710 ± 78
Norepinephrine		290 ± 27	415 ± 123
Epinephrine		173 ± 30	295 ± 57
Femoral vein	22		
Total catecholamine		325 ± 23	545 ± 58
Norepinephrine		234 ± 24	390 ± 51
Epinephrine		92 ± 11	155 ± 35

Values in the table indicate mean $\pm$ SEM. From *Miura* [279]; reprinted with permission.

Miura [279] reported that resting concentrations of plasma norepinephrine in coronary sinus blood in men and women were respectively 20 and 40% greater than the concentrations in aorta and femoral vein blood. On the other hand, the mean resting levels of epinephrine in aorta blood were 30 and 50% greater than those in the coronary sinus and femoral vein, respectively.

The effect of isometric handgrip exercise on hemodynamics and plasma catecholamines is indicated in table V. Both norepinephrine and epinephrine concentrations were elevated by the exercise; the mean increase in total plasma catecholamines was more marked in the coronary sinus (80%) than the aorta (60%) or the femoral vein (68%) [279].

The findings of *Yamaguchi* et al. [430] suggested that the concentration of norepinephrine in coronary sinus blood reliably reflects the activity of the cardiac sympathetic nerves. The results of *Miura* [279] also support this suggestion and they proposed that if plasma catecholamine concentrations are to be used as a reliable guide to the degree of adrenergic activity in the regulation of the heart, it may be necessary to measure coronary sinus blood catecholamines in addition to measuring concentrations elsewhere in the circulation. However, based on calculations of catecholamine outflow (i.e. plasma catecholamine concentration multiplied by plasma flow rate) in the coronary sinus and aorta during rest and exercise, *Miura* [279] reported that very little of the norepinephrine in the general circulation arises from adrenergic innervation and that the notion that cardiac tissue is a major source of circulating catecholamines is untenable.

Additional findings by *Miura* [279] suggest that the difference (delta) between plasma norepinephrine concentrations in coronary sinus and aorta blood is reduced in patients with cardiac dysfunction. A diminished norepinephrine release from cardiac tissue could explain this latter reduction observed in patients with cardiac dysfunction and heart failure. It was further noted that patients with elevated left ventricular end-diastolic pressures (>13 mm Hg) tended to have smaller differences between the concentrations of norepinephrine in coronary sinus and aorta blood than in subjects with relatively normal pressure (<13 mm Hg).

As pointed out by *Miura* [279], this reduced delta norepinephrine in patients with cardiac dysfunction could be due not only to a diminished overflow of the neurotransmitter from the heart into the coronary circulation but also to increased cardiac uptake or enhanced degradation of norepinephrine in the coronary circulation; however, a decreased cardiac uptake of norepinephrine has been found in experimental heart failure. It has been reported that COMT activity is elevated in the failing heart [86], and it has been suggested that this enzyme excess could enhance the catabolism of catecholamines in the coronary circulation and thereby contribute to the reduced difference between concentrations of norepinephrine in coronary sinus and aortic blood of patients with cardiac dysfunction [279].

There is evidence that cardiac concentrations of norepinephrine and tyrosine hydroxylase activity are inversely correlated with pulmonary wedge pressure [86]. Furthermore, there is evidence that the cardiac concentrations of norepinephrine and tyrosine hydroxylase are reduced in heart failure (this subject is discussed in detail in section V. D).

From these latter findings it seems most likely that the reduced difference between concentrations of norepinephrine in coronary sinus and aorta blood observed in heart failure is due to a reduced release of norepinephrine from the sympathetic nerves innervating the heart.

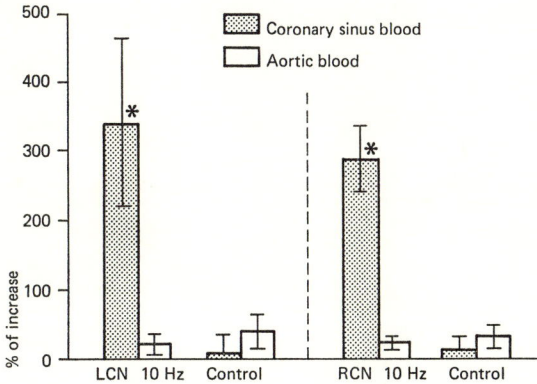

Fig. 11. Catecholamine levels in coronary sinus and aorta following sympathetic nerve stimulation. From *Nadeau* et al. [291]; reprinted with permission.

The effect of electrically stimulating cardiac sympathetic nerves on the norepinephrine concentration in the venous effluent from the left ventricle has been studied by several investigators [239, 291, 430]. *Nadeau* et al. [291] found that catecholamine levels in coronary sinus blood was increased 4-fold by stimulation (at supramaximal intensity for 1 min at 10 Hz) of either the right or left cardio-accelerator nerves, whereas catecholamine levels in aortic blood did not change significantly (fig. 11). Furthermore, it was observed that both coronary sinus blood catecholamine concentrations and physiologic responses of the heart were augmented by increasing the frequency of cardiac sympathetic nerve stimulation; catecholamine levels increased linearly whereas the physiological responses (i.e. heart rate, first derivative of left ventricular pressure, and coronary blood flow) increased in an exponential fashion (fig. 12). Thus, it was concluded that the catecholamine concentration in coronary sinus blood is a reliable index of norepinephrine release during cardiac sympathetic nerve stimulation.

As mentioned earlier, stimulation of the sympathetic cardiac nerves in the dog augment ventricular contraction, heart rate, and arterial pressure. Stimulation of these nerves also influences the atria such that A wave duration is decreased and atrial pulse pressure is augmented [391]. A change in carotid sinus pressure can reflexly (via the central nervous system and cardiac nerves) influence the activity of the heart. A lowering of blood pressure activates the efferent sympathetic nerves and decreases vagal activity; as a result atrial and ventricular contraction are augmented. Increasing

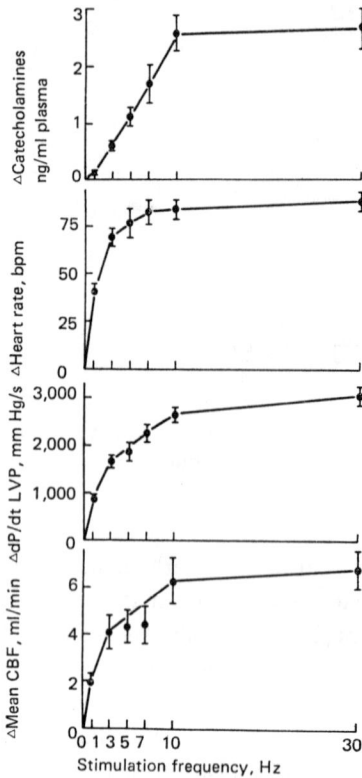

Fig.12. Blood catecholamine levels and physiological response to increasing frequency of sympathetic nerve stimulation. From *Yamaguchi* et al. [430]; reprinted with permission of the American Heart Association.

carotid sinus pressure causes opposite effects [344, 345]. The effects on ventricular ejection dynamics of stimulating sympathetic cardiac nerves are indicated in table VI.

There is evidence that increased sympathetic nerve activity improves synchrony of excitation and contraction within each chamber of the heart. Synchrony is critical to the ability of cardiac ejection to keep pace when the demands for cardiac output increase. Stimulation of the sympathetic cardiac nerves can also cause alterations in the location of the pacemaker. For example, stimulation of the left stellate frequently causes a shift of the pacemaker to points in or around the coronary sinus or AV node [322, pp. 55–56].

Randall [322] has pointed out that it is difficult to depict accurately the precise alterations in the electrocardiogram which may emerge during generalized sympathetic nerve stimulation, presumably because of a differ-

Table VI. Sympathetic influences upon cardiac dynamic events (heart rate kept constant)

Ascending aortic flow
1 Acceleration (increased) (index of contractility)
2 Peak flow (increased, and occurs earlier)
3 Deceleration (increased)   5 Stroke volume (increased)
4 Ejection time (decreased)   6 Cardiac output (increased)

Ventricular pressure
1 Atrial contraction (increased)
2 Max dp/dt (increased) (index of contractility)
3 Min dp/dt (increased) (reflects elevated peripheral resistance)
4 Systolic pressure (increased)
5 Peak systolic pressure occurs later in systole because of the increase in impedance
   (i.e. resistance is increased and the aorta is stiffer)
6 Diastolic pressure (decreased)
7 Diastolic period (increased)

Aortic pressure
1 Upslope (increased)   3 Diastolic pressure (increased)
2 Systolic pressure (increased)   4 Pulse pressure (increased)

Ventricular volume
1 Ejection rate (increased)   4 Diastolic volume (increased?)
2 Ejection volume (increased)   5 Filling rate (increased)
3 Systolic volume (decreased)

From *Randall* [322]; reprinted with permission.

ential nerve distribution to various regions of the myocardium. He has summarized concisely some of the electrocardiographic changes that may be provoked by sympathetic excitation, and has stated that: 'Sympathetic excitation is often accompanied by marked alterations in cardiac rate and rhythm with associated changes in ECG (and His bundle electrocardiogram) morphologies.' Such changes include sinus tachycardia (most commonly during right side stimulations), atrial and ventricular premature systoles, atrial, junctional, and ventricular tachycardias. However, when changes in rate are prevented (as by atrial pacing), reasonably predictable changes may be observed. During left stellate stimulation (LSS) the pacemaker frequently shifts from the SA nodal (or high right atrial) region to junctional (or low atrial) regions with accompanying changes in configuration of the P wave and shortened P-R interval. The RS-T segment is frequently depressed and the terminal phase of the T wave becomes markedly elevated. The Q-T interval may also be significantly prolonged. During right stellate ganglion stimulation (RSS), sinus tachycardia is the most frequently encountered alteration. Because of SA nodal predominance, the P-wave almost invariably remains upright and of normal configuration, appearing to become more prominent in those instances in which the pacemaker was not in the SA node before stimulation. P-R intervals may decrease, particularly in those instances in which the heart rate accelerates, but also in the absence of such acceleration due to increased conduction velocity through the AV node. RS-T segment elevation is sometimes prolonged and is often accompanied by T wave flattening or even inversion. Q-T prolongation may or may not occur [322, p. 57]. *Randall* [322, p. 58] cited evidence that right stellate ganglionectomy or left stellate stimulation prolonged the Q-T interval and increased T wave amplitudes, whereas left stellate ganglionectomy or RSS increased T wave negativity with or without Q-T interval changes. 'These changes could be correlated with changes in ventricular refractory period, right stellate ganglionectomy being associated with refractory period prolongation over the anterior ventricular surface and left ganglionectomy producing prolongation on the posterior surface.' It is pertinent that right stellectomy causes a greater decrease in norepinephrine concentration in the right ventricle and anterior wall of the left ventricle, whereas left stellectomy produces a maximal decrease in the posterolateral areas of the left ventricle [216]. *Randall* [322, p. 58] interpreted alterations in amplitude of the T wave in terms of improved synchrony or greater rate of repolarization of individual segments within the ventricle. He stated that: 'The more synchronous contraction of individual elements of the ventricle during systole, and the more nearly simultaneous relaxation of these elements during diastole imply similar alterations in electrical depolarization and repolarization, respectively.'

It is noteworthy that following excision of the SA node, supraventricular, positive chronotropic effects can be elicited by stimulation of the cardiac sympathetic nerves, thus indicating a regulation of heart rate independent of the SA nodal mechanism.

The effect of sympathetic stimulation on the atrial refractory period remains controversial; however, the ventricular refractory period is clearly shortened. The marked reduction in ventricular fibrillation threshold during stellate stimulation has been attributed to the shortened refractory period induced by sympathetic activation. Ventricular tachycardia can be induced by stimulation of sympathetic nerves, especially those on the left; however, it is unclear whether the sympathetic nerves to the heart play an important role in the genesis of ventricular tachycardias encountered. Although sympathetic activation accompanied by a pronounced increase in mechanical performance of the heart increases ventricular automaticity, it is unlikely that ventricular dysrhythmia would develop except perhaps in the presence of AV block [322, p. 68].

AV conduction time is shortened by sympathetic nerve stimulation; the effect of left stellate stimulation is more pronounced than stimulation on the right – minimal effects occur in the His Purkinje and ventricular muscle conduction. *Randall* [322, p. 63] points out that an important tonic sympathetic influence on AV conduction, combined with an autonomic action on the SA node, indicates a homeostatic role for the sympathetic nervous system on dromotropic, chronotropic and inotropic mechanisms.

There is evidence of cardiotonic influences of the sympathetic nerves; even at rest and under anesthesia nerve impulses at low frequencies can be detected. These tonic discharges change synchronously with arterial pulsation, presumably due to the baroreceptor reflex mechanism.

Sympathetic nerve innervation of the heart can significantly influence the excitability of the myocardium. *Skelton* [361] reported that cardiac sympathectomy in the dog protected against ventricular fibrillation and reduced mortality rate following occlusion of the left circumflex coronary artery from 80 to 0%.

Ebert [99] has shown that cardiac denervation markedly reduced or prevented the release of potassium from the heart after coronary ligation; he concluded that potassium flux was important in the genesis of ventricular fibrillation in the ischemic myocardium.

Coronary occlusion stimulates cardiac afferent fibers [252] which reflexly activate efferent cardiac sympathetic fibers [253] and thereby induce ventricular arrhythmias [141]. Complete cardiac denervation causes a marked depletion of catecholamines in the heart [101, 154]. Despite functional reinnervation the catecholamine concentration in the heart remains markedly reduced [162].

Spurgeon et al. [372] reported the interesting finding that cardiac denervation decreased the norepinephrine concentration in the atria and ventricles to 1–3% of control concentrations whereas epinephrine concentrations were reduced only to 40–60%. The epinephrine concentration in conductile tissue (except for the SA node) did not decrease significantly below control values; however, the norepinephrine concentration in the right and left bundle branches and SA and AV nodal conductile tissue decreased markedly to almost zero. These results indicate that a substantial amount of epinephrine is nonneuronal since denervation is accompanied by degeneration of the sympathetic cardiac innervation with a concomitant disappearance of the norepinephrine-containing vesicles normally present in these nerves. *Hoffman* [180] demonstrated that catecholamines increased diastolic depolarization in Purkinje fibers and pacemaker cells which resulted in increased firing. *Han* et al. [165] also studied the adrenergic effects on ventricular vulnerability and presented evidence that a homogeneous distribution of catecholamines throughout the myocardium tends to reduce vulnerability to arrhythmias.

It is not surprising, therefore, that conditions (e.g. sympathetic nerve stimulation, myocardial ischemia or infarction) which alter sympathetic activity and catecholamine content in a region of the myocardium enhance the chance of developing cardiac dysrhythmias. As pointed out by *Randall* [322, p. 67], such changes, coupled with alterations in recovery of myocardial excitability and disorganized conduction, may lead to multiple reentry circuits and fibrillation. The value of using adrenergic blocking agents, particularly some of the beta blockers, to control and prevent hazardous dysrhythmias, must be considered in a variety of clinical situations afflicting the heart.

It should be mentioned that vulnerability to ventricular fibrillation can be reduced by infusion of a solution of glucose, insulin, and potassium [39]. This protection, which has been demonstrated in both the normal and ischemic canine heart, appears to be partly due to antagonism of adrenergic activity.

Brooks et al. [39] demonstrated that glucose-insulin-potassium solution prevented the reduction in ventricular fibrillation threshold caused by infusing norepinephrine or by electrical stimulation of the left stellate ganglion. They also found that infusion of a glucose-insulin-potassium solution in normal dogs significantly decreased heart rate but plasma concentrations of epinephrine and norepinephrine and mean arterial blood pressure did not change significantly. Furthermore, since the protection against ventricular fibrillation achieved by infusion of glucose-insulin-potassium solution is mediated by extra-adrenergic mechanisms, the latter may, in part, be due to a cardiac membrane-stabilizing action.

Brooks et al. [39] cited the evidence that glucose administration increases cellular uptake of potassium and reduces potassium loss from ischemic myocardium, thereby promoting hyperpolarization and improving conduction. In addition, insulin increases glucose uptake and further enhances hyperpolarization. As a result, electrical stability in ischemic and non-ischemic myocardium is improved [39].

Of interest is a recent report by *Scott* et al. [350] that tyrosine administration decreases vulnerability to ventricular fibrillation in the normal canine heart. One might have anticipated that tyrosine would have increased catecholamine synthesis and peripheral sympathetic activity and thereby increase the vulnerability to ventricular fibrillation. However, tyrosine caused the opposite effect, and it was postulated that this precursor of catecholamines enhances the central catecholaminergic activity in the brain (perhaps as do clonidine and methyldopa), thereby diminishing sympathetic neural outflow to the heart, and *decreases* ventricular fibrillation vulnerability [350]. In support of this postulate, *Scott* et al. [350] cited the evidence that clonidine, which stimulates alpha-noradrenergic receptors in the brainstem, decreases vulnerability to ventricular fibrillation and suppresses digitalis-induced arrhythmias. Furthermore, they cited the evidence that norepinephrine synthesis (and probably its release) and its metabolite (methoxyhydroxy-phenyl glycol) are increased in the brain by administering *L*-tyrosine [350].

C. Parasympathetic-Sympathetic Interaction Involving the Myocardium

A few remarks should be made regarding the parasympathetic control of myocardial function and the interactions of the sympathetic and parat sympathetic innervation of the heart. Recent advances regarding this subjec-have been reviewed by *Levy* [237], and will be considered in this section. It should be appreciated that cholinergic and adrenergic interactions are complex and that their effects on cardiac function may be synergistic or antagonistic. Apparently parasympathetic influences are modulated to some extent by the degree of sympathetic activity. With a high level of sympathetic tone, the vagal center in the medulla is inhibited. Under certain circumstances activation of sympathetics may exaggerate the cardiovascular response to parasympathetic stimulation whereas under other conditions the response is markedly blunted [174]. Elevation of blood pressure can produce reciprocal effects and reduce sympathetic traffic and augment parasympathetic activity, whereas an increase in arterial CO_2 will stimulate both the vagus and sympathetic nerves [9]. For a comprehensive review concerning parasympathetic control of the heart the reader is referred elsewhere [174].

Vagal and sympathetic activity are constantly varying; however, even moderate vagal stimulation can overshadow strong cardiac sympathetic activity and mask sympathetic influences on the myocardium. Furthermore, the response to vagal stimulation is more rapid than that due to sympathetic stimulation. Time dependency of the pacemaker response to vagal stimulation depends partly on high concentrations of acetylcholinesterase in the SA nodal region and consequent rapid hydrolysis of acetylcholine released in this region by vagal terminals; inactivation of norepinephrine released by sympathetic nerve terminals in the same region is less rapid. AV conduction is depressed by vagal impulses whereas the cardiac sympathetic nerves have the opposite effect. Behavior of AV conduction differs from activity in the SA node with respect to autonomic interactions since change in activity of one division of the autonomic nervous system is independent of the background level of the other division.

Vagal stimulation reduces heart rate, automaticity, contractile force and maximal left ventricular dp/dt; however, diminution of contractile force evoked by vagal stimulation or acetylcholine infusion is accentuated if myocardial contractility is first augmented by an infusion of norepinephrine or by cardiac nerve stimulation [185, 374]. *Levy* [237] termed the latter phenomenon 'accentuated antagonism' and has indicated that the principal sympathetic-parasympathetic interactions on the heart can be subdivided into two major categories: reciprocal excitation and accentuated antagonism.

Additional evidence of accentuated antagonism can be demonstrated. For example, immediately following cessation of supramaximal stimulation of the vagi, there is a transient increase in left ventricular pressure caused by a 'rebound' enhancement in myocardial contractility. Also, the greater the background of sympathetic activity the more profound the depressant effect of a given level of vagal activity [237].

Both interneuronal and intracellular processes appear to be involved in the mechanism responsible for accentuated antagonism. Postganglionic cholinergic vagal activity can decrease nerve traffic in cardiac sympathetic nerves and thereby reduce the quantity of norepinephrine released in the heart. This inhibitory effect of acetylcholine on the release of norepinephrine, which has been identified as a muscarinic effect, can be blocked with atropine. On the other hand, nicotonic agents can release norepinephrine in the heart. Many studies have shown that acetylcholine and cholinergic stimulation may release norepinephrine from the heart. *Levy* [237] reported the effect of cardiac sympathetic stimulation, alone and in combination with vagal stimulation, on norepinephrine overflow into the coronary sinus blood (fig. 13). Others demonstrated that superimposition of vagus nerve stimulation on sympathetic stimulation resulted in a frequency-related decrease in coronary sinus catecholamine levels (fig. 14). The inhibitory effect of vagal stimulation on coronary sinus catecholamine overflow can be mimicked by atropine or methacholine (a pure muscarinic agent) and is rapidly reversible (fig. 15).

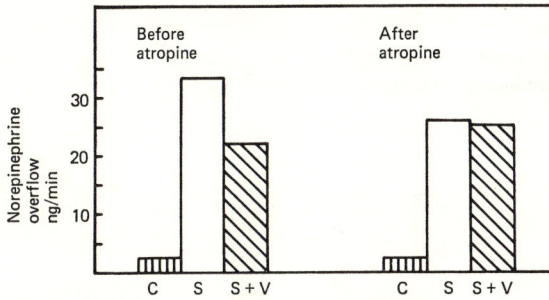

Fig. 13. The rate of norepinephrine overflow into the coronary sinus during the control state (C), during cardiac sympathetic stimulation (S), and during combined sympathetic and vagal stimulation (S + V). The heights of the bars indicate the mean values for 6 anesthetized dogs. The data on the left were obtained before atropine; those on the right after atropine sulfate, 1 mg/kg. From *Levy* [237]; reprinted with permission.

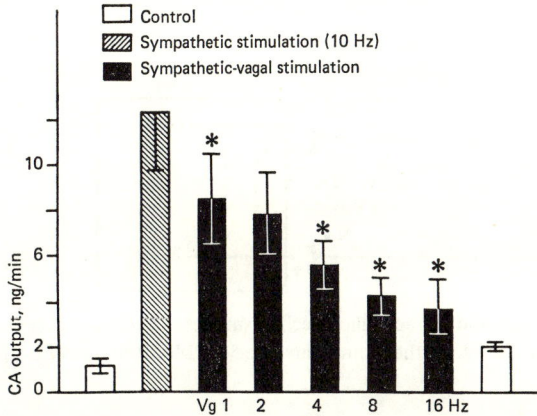

Fig. 14. Effect of sympathetic stimulation with and without parasympathetic superimposition on coronary sinus noradrenaline. From *Lavallée* et al. [230]; reprinted with permission.

Since it is evident that parasympathetic activity can modulate the effects of cardiac sympathetic nerve stimulation, *Nadeau* et al. [291] speculated that this interaction of the autonomic nervous system could explain the depression of ventricular performance caused by vagal stimulation, especially when sympathetic tone is high [239]. They further suggested that such an interaction may explain the protective effect which parasympathetic activity exerts against the development of ventricular fibrillation caused by sympathetic stimulation [291].

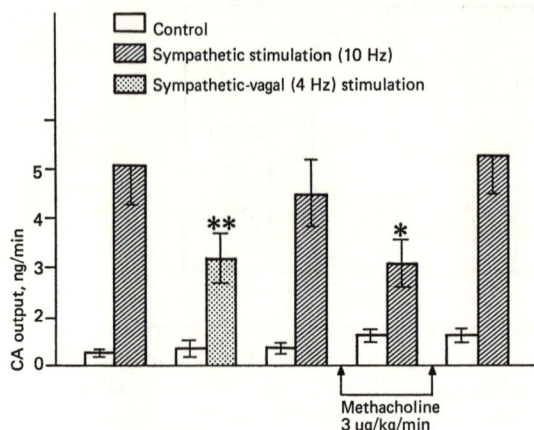

Fig. 15. Reversal of parasympathetic inhibition of coronary sinus blood catechol-amine levels. From *Lavallée* et al. [230]; reprinted with permission.

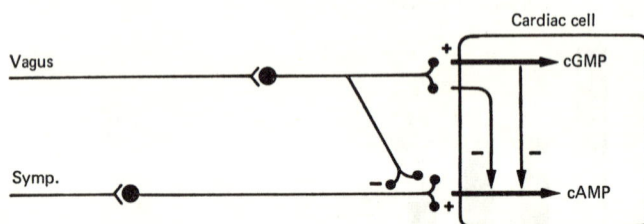

Fig. 16. The interneuronal and intracellular mechanisms responsible for accentuated antagonism between the cardiac sympathetic and vagal nerves. Modified from *Levy* [238]; reprinted with permission.

The mechanism whereby vagal stimulation with acetylcholine secretion diminishes the release of norepinephrine in the heart during sympathetic stimulation remains unclear. Prostaglandins of the E series are released into coronary sinus blood by acetylcholine or vagal activation. Whether or not prostaglandins subserve an intermediary role in suppressing norepinephrine secretion at a given level of cardiac sympathetic activity remains contro-versial [237].

In addition to the muscarinic inhibition of norepinephrine release, accentuated vagal-sympathetic antagonism depends on an intracellular mechanism which reduces the cardiac response to norepinephrine. Acetyl-choline produces a greater reduction in left ventricular contractile force and maximum dp/dt during a simultaneous norepinephrine infusion than when it is given alone. As indicated by *Levy* [237], it appears that cyclic nucleotides are involved in the actions of the sympathetic and parasympathetic nerves on the myocardium and in the interactions between these autonomic nerves. Norepinephrine increases intracellular levels of cyclic AMP and thereby enhances glycogen phosphorylase activity in the myocardium. Acetylcholine has no effect or causes a slight reduction of basal levels of cyclic AMP; yet if cyclic AMP levels are elevated by adrenergic stimulation then acetylcholine profoundly depresses the intracellular level of this cyclic nucleotide. Contrariwise, acetylcholine increases the intracellular level of another cyclic nucleotide, cyclic GMP. It is probable that this latter nucleotide acts as an intermediary by which acetylcholine and cholinergic drugs elicit some of their cardiac effects – responses opposite to those produced by adrenergic stimuli where cyclic AMP is the intermediary. There is evidence that cyclic GMP accelerates hydrolysis of cyclic AMP and may thereby lower the intracellular level of the latter nucleotide [23].

From the foregoing, *Levy* [237] has proposed that vagal stimulation may suppress myocardial cyclic AMP in two ways: (a) by opposing the tendency for adrenergic stimuli to elevate intracellular levels of cyclic AMP and (b) by raising the concentration of cyclic GMP, which may accelerate the hydrolysis of cyclic AMP. Figure 16 schematically depicts the inhibitory influence of terminal vagal fibers on postganglionic sympathetic neurons in the heart; it represents the interneuronal and intracellular mechanisms responsible for accentuated antagonism between cardiac autonomic nerves.

D. Cardiac Reflexes

Reflex mechanisms play a major role in regulating heart function by their influence on autonomic cardiac nerve activity. The following remarks are based mainly on the review of cardiac reflex mechanisms by *Armour* et al. [9].

The central nervous system is continuously monitoring the instant to instant status of the heart via afferent cardiac nerves. In addition to the well-known baroreceptor reflexes, which are so important in autonomic modulation of blood pressure changes, preganglionic sympathetic nerve activity is modulated by other structures, such as receptors in the heart, great vessels and lungs, which respond to mechanical pressure and distortion. Bursts of sympathetic activity appear with each systole whereas sympathetic activity decreases with inspiration. Powerful reflexes arising from cardiopulmonary receptors can activate cardiac sympathetic and parasympathetic nerves simultaneously [9].

Afferent nerves can also be activated by increases in coronary flow and/or pressure, coronary occlusion or myocardial ischemia, and this activation can reflexly cause cardiac sympathetic nerve discharge [41]. Even stimulation of sciatic, brachial and saphenous nerves can evoke excitation of cardiac sympathetic nerves while simultaneously inhibiting vagal activity [188]; these are known as somatosympathetic reflexes. It has been proposed that local cardiac reflexes can activate postganglionic sympathetic nerves via thoracic ganglia [423]. Hence, it is evident that reflex activation of sympathetic efferent cardiac nerves can occur via afferent as well as efferent fibers.

Kezdi et al. [214] observed a decreased cardiac output, blood pressure, heart rate, and a decreased postganglionic sympathetic neural activity following occlusion of the circumflex coronary artery in the dog. Administration of atropine mainly increased the heart rate whereas transection of the vagi increased sympathetic nerve activity, cardiac output, and mean systemic blood pressure. These investigators recommended that cardiogenic shock secondary to myocardial infarction be treated by administering atropine to decrease bradycardia and also by administering isoproterenol and norepinephrine to increase cardiac output and peripheral resistance. However, as pointed out by *Armour* et al. [9], therapy with agents which increase cardiac output and peripheral resistance represent a 'double-edged sword' since they increase cardiac work and coronary blood flow demand.

Hypertension and tachycardia occur in some patients during angina, and vasoconstrictor responses and increased cardiac output have been observed after myocardial infarction as well as following coronary occlusion. Such pressor reflexes may compensate for the decreased contractility of injured myocardium and thus help prevent cardiogenic shock due to myocardial infarction [253]. *James* et al. [197] suggested that serotonin released from platelets in the region of atherosclerotic plaques or infarcts may reflexly cause some of the blood pressure changes associated with myocardial infarction. They observed that injections of small amounts of serotonin into proximal segments of the left coronary artery caused hypertension whereas injection into the distal segments caused hypotension.

In summary, *Armour* et al. [9] have schematically depicted the known reflex connections of the heart and a number of hypothetical pathways (fig. 17). Preganglionic sympathetic fibers leave the spinal cord and make synaptic connections with postganglionic sympathetic cardiac nerves in the

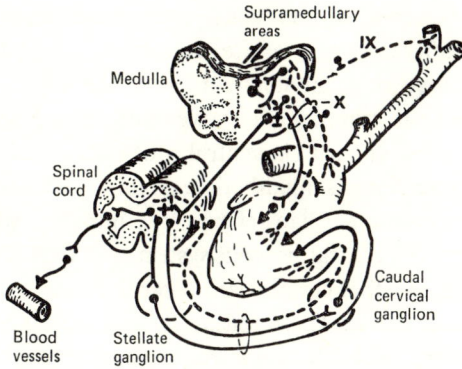

Fig. 17. Diagram of hypothetical reflex connections of the canine heart. Afferent fibers are indicated by dashed lines and efferent fibers by solid lines. Sympathetic cardiac afferent fibers may have excitatory influences on cardiac sympathetic efferent fibers at the spinal level or via supraspinal connections (not shown). Baroreceptor afferent fibers may travel to the medulla from the carotid sinus by way of the glossopharyngeal nerve (IX) or from the aortic arch by way of the vagus nerve (X). These baroreceptors reflexly excite (+) cardiac vagal efferent fibers and inhibit (–) descending spinal sympatho-excitatory pathways to preganglionic neurons which are destined to the heart and blood vessels. Other cardiopulmonary afferents traveling in the vagus nerve may excite or inhibit (±) the descending spinal sympatho-excitatory pathways. Sympathetic cardiac pregangli-onic fibers may synapse with postganglionic neurons in the stellate ganglion or more commonly in the caudal cervical ganglion. From *Armour* et al. [9]; reprinted with permission.

stellate but more often in the caudal ganglia. Preganglionic parasympathetic fibers travel in the vagi and terminate on postganglionic nerves in or near the heart.

Some vagal and sympathetic cardiac efferent nerves show cyclic activity with cardiac rhythm whereas other efferent fibers do not exhibit this activity. The autonomic efferent cardiac nerves are modulated by reflexes which originate in the carotid and aortic baroreceptors. Excitatory impulses are conveyed from these receptors by afferent nerves to the medulla where they inhibit descending reticulo-spinal tracts to preganglionic cell bodies and thereby reduce postganglionic sympathetic outflow to the heart. Usually, when sympathetic activity increases, vagal activity decreases. This reciprocal relationship (indicated in fig. 17) does not always occur since coactivation of both sympathetic and parasympathetic systems may occur under some conditions.

Nonbaroreceptor afferent nerves also influence autonomic nerves to the heart. Cardiopulmonary afferents may excite or inhibit sympathetic outflow to the cardiovascular system. Sympathetic afferent fibers from the heart travel to the spinal cord via sympathetic nerves and dorsal roots; visceral afferents may reach the spinal cord via ventral roots. Stimulation of these afferent nerves primarily activates sympathetic efferents to the heart.

Many cardiac reflexes are cardio-cardiac reflexes, i.e. impulses which arise in the heart, are relayed to the central nervous system, and then return to the heart. Some evidence suggests that certain cardio-cardiac reflexes may occur in extrinsically denervated hearts. These reflexes release norepinephrine in response to weak beats and acetylcholine in response to strong beats [220].

Reflexes from coronary arteries or myocardial infarcts may be depressor or pressor and cause hypotension or hypertension. It is noteworthy that these reflex pathways may be interrupted by coronary bypass surgery. Cardiac reflexes and activation of sympathetic nerves appear to be important in producing several types of arrhythmias. For example, excessive sympathetic activity in humans with the prolonged QT syndrome may cause lethal cardiac arrhythmias [349].

E. Intracranial Mechanism of Cardiac Regulation

Manning [258] has recently reviewed the intracranial mechanisms involved in cardiac regulation. The central nervous system can exert influences on the cardiovascular system. Autonomic activity evoked by diencephalic stimulation can cause a variety of arrhythmias. Electrocardiographic changes which may be associated with intracranial lesions appear related to alterations in the interplay of the sympathetic and parasympathetic nervous systems. Although confusion exists regarding the precise location of bulbar synaptic stations of some cardiac reflexes, efferent nerves which cause vagal cardioinhibitory effects probably reside in the nucleus ambiguus. Hypothalamic pressor areas suppress sinus reflex bradycardia; the limbic forebrain, limbic hindbrain and hypothalamus may be interrelated with baroreceptor reflex pathways. A cortico-hypothalamic sympatho-inhibitory system involves efferent fibers which inhibit sympathetic discharge and stimulate vagal activity.

Manning [258] has indicated that a preeminent vasomotor center in the medulla governing cardiovascular control is too simplistic a concept. He

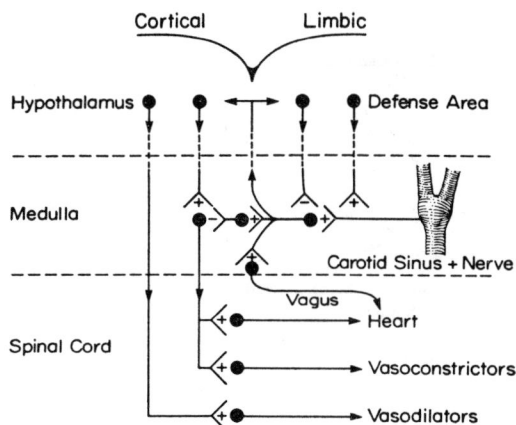

Fig. 18. A schema depicting a hypothalamic-medullary loop whereby descending hypothalamic systems can inhibit components of the carotid sinus medullary input either by presynaptic or postsynaptic mechanisms. Such inhibition permits full development of hypothalamic activity on spinal sympathetic neurons. In addition, an ascending system of the IX nerve engages diencephalic neurons. From *Manning* [258]; reprinted with permission.

states that: 'The addition of a hypothalamic loop to the controller provides the organ with greater flexibility in regulating basic reflex activity.' According to his schema (fig. 18): 'The basic homeostatic reflexes are represented by the IX nerve input which makes synaptic connection in the medullary reticular formation... These medullary synaptic stations are impinged upon by neural systems of diencephalic origin which offer tonic as well as phasic adjustment to cardiomotor and vasomotor spinal outflow.'

F. Spinal Sympathetic Influence on the Heart

For a detailed account of the spinal sympathetic control of the heart the reader should consult the review by *Wurster* [426] on which most of the following discussion is based.

Cardiac changes and alterations in control mechanisms may occur in a number of diseases which involve the spinal cord (e.g. multiple sclerosis,

High lesion

Low lesion

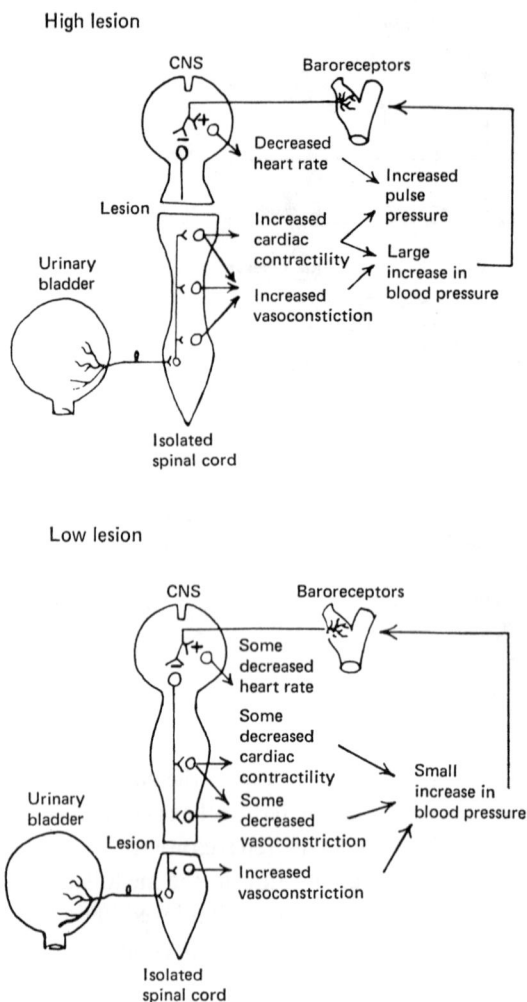

Fig. 19. Diagram of cardiovascular reflexes in response to bladder distension in patients with high (above T_5) and low (below T_5) spinal cord transection. From *Wurster and Randall* [427]; reprinted with permission.

amyotrophic lateral sclerosis, poliomyelitis, tabes dorsalis, Shy-Dragger syndrome, tetanus). Severe tetanus may be accompanied by pronounced hypertension, vasoconstriction, tachycardia and arrhythmias – manifestations resulting from sympathetic hyperactivity [74]. This sympathetic overactivity results from a reduction in mechanisms which normally inhibit sympathetic neurons in the spinal cord [303]. Trauma or mechanical or electrical stimulation of the cord may also markedly affect neural regulation of the heart. For example, increased intracranial pressure can cause marked hypertension by a reflex in which the spinal cord participates [257, 278]. Cardiovascular effects are induced by stimulation of the adrenergic nervous system with release of catecholamines from the sympathetic nerves and adrenal medulla [257].

Spinal animals and patients with lesions above the T_5 level exhibit marked cardiovascular responses during afferent nerve stimulation. Some responses (e.g. marked increase in blood pressure, vasoconstriction, increased cardiac contractility) are due to activation of the adrenergic system; other responses (e.g. bradycardia and vasodilatation of the face, neck, and shoulders) result from reflex activation of the vagus which is induced by the hypertension [127, 257]. From studies of patients with spinal cord lesions, *Wurster and Randall* [427] concluded that in high spinal lesions (T_5 or above) afferent nerve stimulation (by bladder distension) caused vasoconstriction and also increased cardiac contractility and elevated mean and pulse pressure. The latter indicated inotropic effects of the sympathetic cardiac innervation. In patients with lesions below T_5, marked increases in pulse pressure did not occur (fig. 19).

A number of investigators have reported spinal reflexes which especially involve the heart. In spinal animals, increased coronary artery pressure, occlusion of the coronary sinus, and myocardial ischemia cause increased sympathetic activity accompanied by cardiac arrhythmias. These cardio-cardiac reflexes can be alleviated by sectioning T_1 to T_5 dorsal roots [426].

There is evidence that the sympathetic nervous system to the heart and vascular system may be differentially controlled, i.e. activation or inhibition of the sympathetic nerves to one organ or area can occur without a similar response to others. As stated by *Wurster* [426], 'The monolithic view of mass sympathetic discharge has fallen.'

Somatic and visceral afferent nerves can excite and inhibit postganglionic nerves to the heart and blood vessels. The afferent excitatory and inhibitory impulses ascend to the brain in pathways located respectively in the dorsolateral sulcus and dorsolateral funiculus regions of the spinal cord; excitatory fibers activate pressor pathways in the reticular formation of the brainstem which descend in the dorsolateral funiculus. The inhibitory afferents may suppress descending excitatory pathways and activate descending inhibitory pathways [426].

G. Autonomic Control of Nodal Tissue

Urthaler and James [393] have reviewed the cholinergic and adrenergic control of the sinus node and atrioventricular (AV) junction. Formation, integration, and conduction of cardiac impulses all occur in specialized myocardial cells which are profoundly influenced by autonomic cardiac nerves and a variety of blood-borne substances, including the catecholamines. The normal pacemaker is the sinus node, and resting heart rate results from synchronous impulses generated spontaneously from these cells. If sinus node activity fails, alternate pacemakers become available – the major substitute being in the region of the AV node and His bundle. Except for the sinus node, the AV junction has the richest adrenergic and cholinergic nerve supply in the heart. Cholinesterase is also heavily concentrated in these areas and in the bundle of His [393]. *Urthaler* et al. [394, 395] have shown that there is a mathematical relationship between automaticity in the sinus node and in the AV junction.

The sinus node is exposed to norepinephrine which is released as the neurotransmitter from the cardiac sympathetic nerves; it is also exposed to catecholamines which are released elsewhere (i.e. epinephrine and nor-epinephrine from the adrenal medulla and norepinephrine from other sympathetic nerves) and reach the heart through the blood. Perfusion of the sinus node with epinephrine or norepinephrine mimics adrenergic stimulation of the node (e.g. by electrical stimulation of the right stellate ganglion). Immediate sinus tachycardia occurs on perfusing with epinephrine, nor-epinephrine or isoproterenol; however, isoproterenol hydrochloride is about ten times more potent (on a weight basis) than the naturally occurring catecholamines. Following administration of maximally effective doses of these agents, tachycardia rarely exceeds 3 min [393]. Dopamine also causes an immediate tachycardia but is less potent by weight than norepinephrine. On the other hand, *L*-dopa (the precursor of dopamine) produces a pro-longed tachycardia which is slow in onset and less marked than that produced by norepinephrine. The effect of *L*-dopa may be due to synthesis of dopamine and norepinephrine. Neither tyrosine nor phenylalanine exert any chrono-tropic effect on perfusion [196].

A positive chronotropic effect can result from (1) a direct accelerating effect of an agent on sinus node cells; (2) release of intranodal catecholamine stores; (3) inhibition of reuptake mechanism for catecholamines in the heart; (4) inhibition of catabolism of the catecholamines; or (5) an anticholinergic action on the vagus or on the local response to acetylcholine. Negative chronotropic responses are chiefly augmented cholinergic effects; however, bradycardia may be mediated through antiadrenergic mechanisms (e.g. blockade of beta adrenergic receptors, local norepinephrine depletion, or neural inhibition of norepinephrine release [393].

It appears that norepinephrine in the sinus node exists in at least two pools – that released by neural stimulation and that which can be released by tyramine [196].

Perfusion of the AV junction with epinephrine, norepinephrine or isoproterenol causes a junctional tachycardia which is almost immediate in onset and of brief duration. Perfusion with tyramine or quanethidine causes a release of norepinephrine and produces an immediate onset of tachycardia; however, the tachycardia is prolonged – similar to that seen following sinal node perfusion with these agents.

Beta-adrenergic blockade of the sinus node of the dog diminishes heart rate by about 20%, whereas blockade of the AV junction reduces the rate by 45%.

Positive and negative chronotropic effects of certain drugs can be mediated by mechanisms independent of the adrenergic and cholinergic systems. For example, perfusion of the sinus node or AV junction with glucagon or guanosine will produce tachycardia despite beta-receptor blockade or pretreatment with drugs which block adrenergic neural transmission. Also, theophylline and aminophylline have chronotropic and inotropic effects which are not abolished by beta-receptor blockade. Adenosine perfusion of the sinus node or AV junction causes a profound bradycardia despite the presence of a cholinergic muscarinic blockade; bradycardia is not altered by beta-receptor blockade, catecholamine depletion or sympathetic nerve blockade [393]. Although there is some evidence that cyclic AMP may be the mediator for the chronotropic and inotropic effects of the catecholamines [380], the mechanisms whereby the catecholamines and other substances cause changes in heart rate and contractile force remain uncertain. It is noteworthy that perfusion of the sinus node with cyclic AMP produces only a bradycardia which is unaltered by pretreatment with atropine or a beta-receptor blocker [194].

H. Autonomic Influence on the Coronary Circulation

The adrenergic receptors and their influence on coronary blood flow have been discussed earlier in this chapter. In this section the autonomic control of the coronary circulation will be considered in more detail. The following is based primarily on the recent review of autonomic control of the coronary circulation by *Pace* [304].

Metabolic activity of the heart is crucial in the control of its blood supply. Coronary blood flow is closely related to myocardial oxygen consumption. Recent evidence suggests that adenosine, which is released from myocardial cells when oxygen demand is increased, may play a key role in causing vasodilatation and thus augmenting coronary blood flow and oxygen delivery. Despite the central role exerted by metabolic factors in controlling myocardial perfusion [335], there is no doubt that the autonomic nervous system also influences coronary blood flow. Beta-adrenergic blockade diminishes vasoconstriction resulting from cardiac sympathetic nerve activity; parasympathetic stimulation causes coronary vasodilatation [304].

Sympathetic nerves have close contact with the muscular layer of arteries in the heart, although these nerves do not appear to penetrate the media to any extent [103]. There is evidence for innervation of coronary arteries by sympathetic and parasympathetic nerves, whereas capillaries appear uninnervated. The media of large coronary vessels are not in close approximation with nerve terminals and it has been suggested that these vessels may only be under moderate or mild control by the autonomic system. In contrast, nerves become more numerous and in closer association with smooth muscle as the coronary vessels become smaller – probably indicating a greater autonomic control of the coronary vessels as they become smaller. Since the neurotransmitter released at these nerve terminals has to diffuse a certain distance before it reaches the media, it is possible that only the outer cells of the muscular wall are stimulated to contract while cells near the lumen remain unaffected [304].

As mentioned previously, the cardiodynamic metabolic effects dominate the coronary blood flow response during sympathetic nerve stimulation. *Granata* et al. [155] found that immediately after the onset of electrical stimulation, vasoconstriction occurred and was accompanied by a decreased coronary flow while heart rate and blood pressure remained unchanged; with subsequent augmentation of heart rate and contractility, the vasoconstriction was converted to vasodilatation.

In some species immediate coronary vasodilatation may be induced by electrical stimulation of sympathetic nerves to the heart. In the cat, such vasodilatation can be abolished and converted to vasoconstriction by beta-adrenergic blocking agents [40]. This latter finding suggests that immediate coronary vasodilation during sympathetic stimulation is due to a beta-adrenergic effect. Furthermore, the vasoconstriction (which occurs following beta-adrenergic blockade) can, in turn, be abolished by alpha-adrenergic blockade with dibenzyline. It appears, therefore, that during sympathetic stimulation alpha-adrenergic vasoconstrictor activity becomes manifest if coronary vasodilatation and cardiac metabolic effects are prevented by beta-adrenergic blockade with propranolol. If only the metabolic effects are prevented by administration of practolol (which selectively blocks only the metabolic effects on the heart caused by sympathetic stimulation), then stimulation of cardiac sympathetic nerves will not significantly diminish coronary blood flow. The latter finding suggests that alpha-adrenergic constrictor effects may be inhibited in the presence of a beta-adrenergic dilator system [293, 304].

It is noteworthy that beta-adrenergic blockade in the anesthetized dog results in a significant decrease in myocardial oxygen consumption and an increase in cardiac efficiency [293]; however, during beta blockade, cardiac sympathetic nerve stimulation causes coronary artery vasoconstriction and a mean reduction in coronary sinus blood PO_2 of 6 mm Hg. The latter effect can be prevented by alpha-adrenergic blockade [119, 120]. *Feigl* has suggested that coronary alpha-adrenergic receptors may play a role in controlling myocardial oxygen consumption by retarding microvascular flow – thereby permitting increased oxygen extraction by the heart. Others have presented evidence that the coronary vasoconstriction due to alpha-receptor activity may increase nutritive flow to myocardial tissue [200].

Studies by *Gregg* et al. [157] revealed that cardiac denervation reduced the mean circumflex coronary blood flow in anesthetized dogs to about 50% of that in the innervated heart; resting cardiac oxygen consumption also decreased in a manner parallel to the reduction in coronary flow. Following cardiac denervation, coronary blood flow and myocardial function appear adequate even during moderate stress; however, cardiac sympathetic innervation may be of value for optimal control of blood flow and myocardial performance under conditions of mild to moderate stress. Certainly during severe stress sympathetic innervation of the heart is mandatory for maximal performance of the heart (see discussion in section IV.I).

In vivo studies, therefore, support the concept that both beta- and alpha-adrenergic receptors are present in the coronary arteries; under appropriate conditions such as cited above, beta-receptor stimulation produces smooth muscle relaxation and vasodilatation whereas alpha-receptor stimulation causes muscle contraction and vasoconstriction.

Furthermore, as mentioned previously (section III.B), in vitro studies revealed that catecholamines caused segments of small coronary arteries (250–500 μM) to relax whereas large arteries (1.5–2.5 mm) were caused to contract. In small arteries, relaxation responses were abolished by beta-adrenergic blockers and often converted to contraction, whereas contraction responses in large arteries were abolished by alpha-adrenergic blockers [440].

It should be mentioned that alpha- and beta-adrenergic receptors have been pharmacologically identified in coronary arteries of the unanesthetized dog. *Pitt* et al. [311] found only coronary dilatation during the initial 20–30 s following intracoronary infusion of norepinephrine. *Vatner* et al. [403] demonstrated that intravenous administration of norepinephrine resulted in a decrease in coronary vascular resistance for the initial 20–30 s followed by a prolonged coronary vasoconstriction. This constrictor response occurred despite increases in left ventricular pressure and contractility. Failure of previous investigators to demonstrate the latter response probably resulted from the effects of anesthesia on coronary vascular reactivity [401]. When propranolol was administered to unanesthetized dogs, beta-adrenergic activity appeared to be minimal since no significant alterations in coronary or systemic hemodynamics occurred [312].

It is noteworthy that although some forms of stress, such as hypoxic hypoxia, may enhance beta-adrenergic activity, the latter is not essential for the coronary vasodilatation which occurs [109] under conditions of hypoxic stress.

With regard to reflex control of the coronary circulation, it appears that bilateral carotid occlusion following bilateral vagotomy evokes an increased heart rate, blood pressure, and circumflex coronary artery blood flow. *Feigl* [118] found that a marked decrease in coronary artery resistance accompanied these changes. In the presence of beta blockade, diastolic resistance in the coronary arteries increased; however, bilateral upper thoracic sympathectomy abolished this rise in resistance. The results suggest that alpha-adrenergic coronary vasoconstriction may participate in the reflex induced by carotid sinus hypotension [119]. Conversely, simulation of carotid sinus hypertension (by electrically stimulating the carotid sinus nerves) in the conscious dog resulted in a decreased mean and late coronary resistance, heart rate and aortic pressure. Neither beta blockade nor atropine administration altered the coronary dilator response; however, the response was prevented by alpha-receptor blockade [402]. Apparently, carotid sinus nerve stimulation results in a reduction in the resting sympathetic constrictor tone [1, 304].

Carotid chemoreceptor stimulation by intracarotid administration of nicotine causes a decrease in coronary vascular resistance and an increase in coronary blood flow. Studies by *Vatner and McRitchie* [405] in conscious dogs indicate that this coronary vasodilatation induced by nicotine is caused to a minor degree by reflex activation of vagal cholinergic fibers; however, the major cause appeared to be withdrawal of alpha-adrenergic constrictor tone which seemed related to the onset of hyperventilation initiated by chemoreceptor stimulation.

Finally it is noteworthy that activation of intracardiac receptors may result in reflex coronary dilatation through a vagal pathway [120].

It has been speculated that activation of alpha-adrenergic receptors (which occurs in the presence of beta blockade during baroreceptor hypotension) may be responsible for precipitating anginal pain in some patients with coronary insufficiency who are being treated with beta-blocking drugs [304].

Pace [304] has summarized points of importance regarding autonomic control of the coronary circulation: Synaptic association of sympathetic and parasympathetic nerves becomes closer as the arteriolar level is approached – thus providing more control of the smaller vessels which are especially involved in nutritive blood flow. Vagal cholinergic excitation produces coronary vasodilatation whereas adrenergic excitation causes either vasoconstriction or vasodilatation. Alpha-adrenergic constriction can only be demonstrated after beta blockade, which prevents both the metabolic cardiac effects and the coronary vasodilatation caused by sympathetic stimulation. Therefore, as *Pace* [304] has pointed out, it is difficult to evaluate the functional significance of the alpha receptors in the presence of augmented myocardial metabolism which is associated with the release of metabolites causing vasodilatation.

Pace [304] has concisely stated that: 'The coronary circulation, therefore, appears to possess two possible but relatively weak mechanisms for vasodilatation, e.g. cholinergic parasympathetic and beta-adrenergic sympathetic. Based on preliminary evidence it may be surmised that each may participate selectively under different physiological conditions. It is probably fair to assume that active coronary vasodilatation via beta-adrenergic mechanisms accompanies generalized myocardial augmentation via the sympathetic nerves, enhancing the subsequent coronary vasodilatation. Conversely, active cholinergic dilatation may arise as a sequela from reflex activation of vagal cholinergic fibers to the myocardium without specific vasomotor regulator functions. A third physiologic mechanism potentially responsible for coronary vasodilatation is the inhibition of tonic vasoconstrictor activity in the alpha-receptor system. Thus, while evidence may be marshalled for each of at least three neurogenic mechanisms for regulation, it is probable that all are subsidiary to the direct action of adenosine, or comparable metabolites, acting in consort with varying levels of blood oxygen during widely varying metabolic demands upon the myocardium.'

I. Cardiac Denervation and Reinnervation

Chemical denervation of the heart by the use of atropine with beta-adrenergic blockade, reserpine, 6-hydroxydopamine or immunosympathectomy lacks specificity, since it affects the autonomic nervous system in areas other than the heart. On the other hand, surgical denervation of the heart provides a more precise means of evaluating the role of cardiac innervation on myocardial performance at rest and during stress; however, it has been emphasized that the method of denervation can radically influence results. The denervation must be complete but without disturbance of other viscera (e.g. the lungs and gastrointestinal tract). The effect of anesthesia on cardiovascular responses and the fact that reinnervation may occur as early as 20–30 days following cardiac denervation in the subhuman species must be appreciated [89, 209]. An understanding of the effects of complete or partial cardiac denervation is important with regard to its clinical implications in heart transplantation [355, 377], in autonomic neuropathies which may accompany diabetes mellitus [242] and in Chagas' disease [6]. Apparently diabetic autonomic neuropathy can result in complete cardiac denervation with a fixed heart rate and a heart that is unresponsive to sympathetic or parasympathetic stimuli [242].

It is noteworthy that studies in a large series of humans with cardiac transplants revealed that the physiologic function of the heart was normal, although the response of the heart rate to exercise was somewhat reduced and required a 'warm-up' period [378]. None of these patients had any evidence of cardiac sympathetic or parasympathetic reinnervation, as indicated by fulfillment of the criteria listed in table VII [170]. Infusion studies with norepinephrine, isoproterenol and propranolol demonstrated that the beta-adrenergic receptors in the conduction system of the denervated heart were intact since they responded to these agents in a manner indistinguishable from the cardiac response of normal subjects [47]. Hence, the conduction system in the denervated heart may respond to circulating catecholamines [101, 325]. A discussion of autonomic denervation and arrhythmias in human

Table VII. Evidence for cardiac denervation

---

1 Donor atrial rate greater than recipient atrial rate
2 Absence of cardiac pain with ischemia and infarction
3 Time course of heart rate rise with exercise delayed
4 Failure of donor sinus node to respond to atropine
5 No heart rate increase with tyramine (60 µg/kg)
6 No heart rate increase with blood pressure drop due to amyl nitrate
7 No histologic reinnervation (limited neural elements)

From *Harrison and Mason* [170]; reprinted with permission.

---

cardiac transplants appears in section V. For a thorough account of denervation and reinnervation of the heart, the reader is referred to the review by *Kaye* [209].

Donald, Shepherd, and associates performed a series of studies which clarified the functional ability of the denervated heart and the influence of sympathetic innervation on myocardial performance [89–93]. Total cardiac denervation in the mongrel dog resulted in an absence of sinus arrhythmia and a very regular R–R interval.

The heart rates were between 90 and 120 bpm (quite similar to rates in the normal dog) if the animal was not disturbed. At rest, values for cardiac output, stroke volume, left ventricular systolic pressure, left ventricular dp/dt and left ventricular maximal ejection rate were similar to values in normal dogs. The immediate and marked tachycardia observed in normal dogs that were startled or exposed to emotional excitement was absent in dogs with cardiac denervation. Intravenous administration of atropine or tyramine to denervated dogs did not cause a change in heart rate, in contrast to the tachycardia seen in normal dogs. Intravenous administration of catecholamines induced a more pronounced elevation of heart rate in the denervated than intact dog. The denervated heart was more responsive to norepinephrine than epinephrine.

Mild exercise caused a modest tachycardia and an increased cardiac output; however, the latter was almost entirely due to an increased stroke volume which accompanied the increased left ventricular end-diastolic pressure and the increased fiber length. (Similar hemodynamic alterations have been observed in man following cardiac transplantation.) On the other hand, the increased cardiac output in the normal dog during exercise resulted almost entirely from the increased heart rate. With severe exercise, the heart rate increased slowly and reached a much reduced peak value during the first 2 min of running as compared to the peak value in the normal dog. The pattern of oxygen uptake during and immediately following exercise was similar in the normal and in the dog with cardiac denervation; oxygen uptake and cardiac output were as high before as after denervation. These investigators also demonstrated that even with the extreme effort of racing performed by greyhounds (which probably required some anaerobic metabolism) there was only a very minor reduction in the ability of the denervated heart to meet the circulatory demands of maximal exercise [91].

Bilateral adrenalectomy in mongrel dogs maintained on cortisone replacement did not alter the response of the denervated heart to exercise; therefore, since the technique of total cardiac denervation depletes myocardial catecholamines [71], *Donald and Shepherd* [93] concluded that the tachycardia observed in their denervated animals could not be ascribed to catecholamines released from the heart or adrenal glands. Furthermore, the increase in cardiac rate was not dependent on changes in right atrial transmural pressure or blood temperature.

Subsequently it was shown that moderate exercise of dogs with chronic cardiac denervation liberated a substance (presumably catecholamines) into the blood which in turn accelerated the heart rate of an isolated denervated heart. This acceleration could be prevented by beta-adrenergic blockade with propranolol. On the other hand, it was found that beta-adrenergic blockade with a similar dose of propranolol administered to mongrel dogs with chronic cardiac denervation did not alter the response of the heart rate to moderate exercise [160]. These findings indicated that circulating catecholamines were not responsible for the tachycardia accompanying moderate exercise in dogs with chronic cardiac denervation [92]. From the foregoing it was concluded that exercise tachycardia in dogs with chronic denervation resulted from an intrinsic property of the heart rather than from a blood-borne agent.

As demonstrated by *Donald* et al. [91], there was very little reduction in myocardial performance in the dog with cardiac denervation, even with severe exercise. (Supersensitivity of the denervated heart to circulating catecholamines [93] may be an important factor in augmenting performance of the denervated heart during intense exercise [160]). Furthermore, beta-adrenergic blockade in the normal greyhound only slightly increased racing time and maximal heart rate (apparently beta adrenergic blockade abolishes the effects of circulating catecholamines on the heart but only partially blocks the reflex stimulation of cardiac sympathetic nerves evoked by exercise [90]). However, in the presence of both cardiac denervation and beta adrenergic blockage, the capability of performing severe work [92] or racing [90] was markedly curtailed. Most of the mongrel dogs failed to complete a heavy work load and in the greyhounds cardiac acceleration was markedly limited, racing time prolonged, and animals finished in a state of collapse. *Donald* and associates concluded that the cardiostimulant action of both the sympathetic innervation of the heart and circulating catecholamines was necessary for maximal myocardial performance. Interruption of one of these mechanisms (by cardiac denervation or beta-adrenergic blockade) reduced maximal performance only slightly; however, withdrawal of both support mechanisms severely limited performance of maximal exercise. In the presence of cardiac denervation and beta-adrenergic blockade, the heart must depend solely on the Frank-Starling length-tension mechanism to increase its power of contraction. The latter mechanism is sufficient to handle a moderate work load but incapable of sustaining adequate myocardial function during maximal exercise [90]. It is noteworthy that similar hemodynamic alterations have been observed during mild exercise in man following cardiac transplantation [355, 377].

Studies on the metabolism of the denervated heart have been few and controversial. Although there appear to be no major alterations in myocardial metabolism, *Gregg* et al. [157] found that in denervated conscious dogs at rest and during exercise the coronary blood flow and myocardial oxygen consumption were about half that of normal dogs under similar conditions.

In summary, *Donald* [89] stated that: 'Animal studies and later clinical experience seem to have satisfied that initial question of the competence of the transplanted heart to meet the demands of everyday activity. The heart deprived of extrinsic cardiac nerves adequately meets the demand of pressure and volume loading by the length-tension mechanism and a limited intrinsic tachycardia or, if the stress is sufficiently severe, by the additional excitatory effect of circulating catecholamines.'

As previously mentioned, the myocardial content of catecholamines and the sympathetic nerves to the heart appear to be involved in the genesis of arrhythmias and ventricular fibrillation which develop following acute occlusion of the coronary artery in the normal dog. Arrhythmias and ventricular fibrillation induced by coronary occlusion or a potassium releasing agent in the normal dog can be prevented by cardiac denervation [100, 101, 347, 398].

Other studies involving the stress of hypertension, hypotension, anemia or hypoxia in animals and man with cardiac denervation indicated that the neurally induced tachycardia is an important part of the cardiac response to stress; the denervated or transplanted heart responds less than normally to these forms of stress [209]. Studies on the transplanted human heart after beta-adrenergic blockade suggest that the heart has an intrinsic ability to increase its rate in addition to the acceleration caused by circulating catecholamines [355]. Studies also indicate that the denervated or transplanted heart becomes almost totally deplete of catecholamines [71, 73, 154] and that it obeys *Cannon's* [50] 'laws of denervation' and becomes supersensitive to circulating catecholamines [72, 85, 92, 93, 100, 316]. As a result the denervated or transplanted heart responds with increases in heart rate, force of contraction and cardiac output to infusions of even minute concentrations of catecholamines. Supersensitivity may be explained by a loss of intra-neuronal binding of circulating catecholamines [85]. Evidently, the ability of the transplanted heart to synthesize, bind and store catecholamines is dependent on its content of sympathetic nerves [315].

Uptake, retention, and synthesis of catecholamines by the transplanted heart was found to be only a few percent of that in the normal intact heart; however, little change occurred in the in vitro activity of the enzymes (mono-amine oxidase and catechol-*O*-methyl transferase) which metabolize the catecholamines. Since catecholamines coming in contact with the myo-cardium of the transplanted heart are not taken up by neurons, they are metabolized largely by COMT as are the circulating catecholamines [315].

Recent investigations by *Palmer* et al. [306] indicated that denervation causes changes in the postjunctional adrenergic receptor as well as the pre-junctional uptake mechanism. Since adenylate cyclase activity from the atria and ventricles of denervated hearts exhibited a 2- to 3-fold greater increase in response to norepinephrine than normal hearts, it was suggested that denervation supersensitivity might depend on both prejunctional and post-junctional changes.

The response of the denervated heart to sympathomimetic drugs (e.g. tyramine, metaraminol, ephedrine and mephentermine) which normally displace and release catecholamines is attenuated, whereas the responses to ouabain, calcium chloride, and glucagon are similar to those in the normal heart [209].

Reinnervation of the heart may occur in the subhuman species but not in the human after cardiac transplantation. Functional responses to reflex and electrical stimulation of the nervous system return to normal within 1–3 years following cardiac denervation [209]. It is noteworthy that functional reinnervation of various segments of the heart, the return of reflexes, and decrease in supersensitivity of segments of the myocardium have been correlated with the return of local myocardial catecholamine content [209]. *Kaye* [209] has demonstrated that the atria (particularly the left atrium) recover their norepinephrine first and that subsequently it reappears in the base of the ventricles and finally in the apical area. In his experience, sympathetic reinnervation of the heart occurred in a base-to-apex direction and was complete about 9 months after denervation. Although return of function (which appeared complete by 9–12 months) parallels return of measurable quantities of myocardial norepinephrine, it is interesting that concentrations of norepinephrine were only a fraction of normal concentrations even 26 months after denervation [209]. (Normal concentrations are greater than 2 µg/g for atrial tissue and 1 µg/g for ventricular tissue.)

J. Neural Influences on Cardiac Electrical Activity

Not only do the autonomic nerves to the heart alter mechanical and metabolic function, they also influence electrical activity. All these effects are caused by actions of norepinephrine and acetylcholine, both of which alter sarcolemmal ionic conductances and thereby produce changes in membrane properties and transmembrane potential.

Norepinephrine may induce slight hyperpolarization which is probably caused either by an activation of the sodium pump or a change in the sodium: potassium ratio. If an action potential is induced by sympathetic nerve activation, it reduces transmembrane potential and augments an influx of calcium and probably some other ions. Calcium is the essential stimulus for excitation-contraction coupling and is also important in regulating intracellular calcium stores [181].

Norepinephrine alters the action potential duration to varying degrees. The explanation for this variability is unclear. However, there is evidence for the existence of both beta- and alpha-adrenergic receptors: the former increase and the latter decrease atrial and idioventricular pacemaker rate [333]. Activation of beta receptors causes action potential shortening, whereas activation of alpha receptors causes prolongation [142]. The magnitude of the change is small and masked by rate-dependent changes in action potential duration, since catecholamines usually accelerate heart rate [181]. The actions of catecholamines on beta- and alpha-cardiac adrenergic receptors, as summarized by *Reder and Rosen* [326] appear in table VIII.

Catecholamines exert an important effect on the myocardium by increasing automaticity, i.e. they increase rate of spontaneous firing of spe-

Table VIII. Actions of catecholamines on alpha and beta cardiac adrenergic receptors[a]

Beta effects

An increase in the sinus rate

Increased conduction velocity through the AV node (thereby resulting in shortening of the A-H interval in His-bundle electrograms and the P-R interval in ECGs)

A decrease in both the functional and effective refractory periods of the AV node, permitting an increase in ventricular rate during AV block due to a decrease in the degree of block

A slight shortening of the relative refractory period of the ventricular specialized conducting system

Changes in ventricular vulnerability to arrhythmias, as reflected in a decreased ventricular fibrillation threshold and altered Q-T intervals in the ECG

Alpha effects

Depression of spontaneous automaticity of Purkinje fibers and atrial specialized conducting fibers

An increase in cardiac action-potential duration

A decrease in potassium uptake by Purkinje fibers

Conduction block in partially depolarized Purkinje fibers

---

[a] Table compiled from data of *Reder and Rosen* [326].
From *Manger* [254]; reprinted with permission.

---

cialized cardiac fibers. Studies on cardiac Purkinje fibers indicate that catecholamines increase the slope of the slow diastolic phase of depolarization (phase 4) so that the transmembrane potential reaches threshold more rapidly. This effect occurs consistently with relatively high concentrations of norepinephrine; however, with very low concentrations of this catecholamine, a slight decrease in the slope of phase 4 invariably occurs which diminishes the rate of autonomic firing. Extremely high concentrations of epinephrine may cause incomplete repolarization and a series of oscillatory depolarizations before repolarization is finally complete. It appears that a slowing of automaticity is due to stimulation of alpha-adrenergic receptors whereas an acceleration is caused by beta-receptor activation [181]. Recently it has been shown that the effects of beta-adrenergic amines on automaticity are greater in neonates than in adults [265, 333]. Figure 20 reveals modifications of the transmembrane action potential recorded from the sinoatrial node which result in altered automaticity.

Experimental evidence suggests that the effects of catecholamines on transmembrane potentials may result from activation of adenylate cyclase and the generation of cyclic AMP. Exposure of Purkinje fibers to dibutyryl

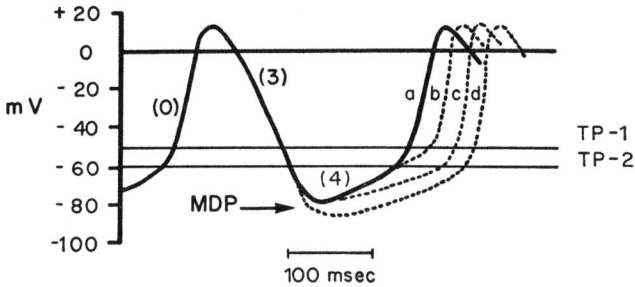

Fig. 20. Diagrammatic representation of the transmembrane potentials of a fiber in the sinoatrial node showing the changes responsible for alterations in the normal automatic mechanism. MDP = Maximum diastolic potential: (4) indicates phase 4 of the transmembrane potential; TP = threshold potential. The diagram indicates that a decrease in the slope of phase 4 (upstrokes a to c), a shift of threshold potential from TP-2 to TP-1 (upstrokes a to b) or an increase in maximum diastolic potential (from c to d) all can decrease the rate of automatic firing. From *Hoffman* [181]; reprinted with permission.

cyclic AMP produces effects similar to those caused by the catecholamines on phases 2, 3 and 4 of the transmembrane potential. Furthermore, intracellular injection of cyclic AMP mimics the effects of catecholamines on phase 4 and shortens the action potential [389].

   The effect of stimulating beta-adrenergic receptors in various cells of the heart varies. Activation of beta receptors of cells in the sinus node augments the slope of phase 4 depolarization and thereby increases the frequency with which they generate action potentials. Exogenous catecholamines and sympathetic nerve stimulation may also shift the pacemaker site within the sinus node. Stimulation of beta receptors of atrial cells has little effect on action potentials in the working myocardium, although high catecholamine concentrations accelerate repolarization of normal human atrial cells. Specialized cells in the crista terminalis of the atrium, however, respond to beta-adrenergic stimulation by the appearance or enhancement of spontaneous diastolic depolarization.

   The mitral valve leaflets are richly innervated with sympathetic fibers; catecholamines increase pacemaker activity in these leaflets but by a mechanism different from that in the atrium or sinus node. Catecholamines initiate a delayed after-depolarization which responds to an increased rate of stimulation by an increase in spontaneous action potentials.

   Beta-adrenergic stimulation of AV cells improves and accelerates conduction through the AV node. Catecholamines increase the rates of depolarization in cells of the upper (AN) and middle (N) region of the AV node.

Table IX. Effects of catecholamines on the electrophysiological properties of the heart

Region	Physiological response	Effects on action potential
SA node	acceleration of pacemaker, shortened refractory period	accelerated diastolic depolarization[a], accelerated repolarization[b] (hyperpolarization)[b]
Atrial myocardium	shortened refractory period (enhanced contractility)[a]	accelerated repolarization[b]
AV node	accelerated conduction	increased amplitude[a] increased rate of depolarization[a]
His-Purkinje system	promotion of pacemaker activity, shortened refractory period	accelerated diastolic depolarization[c], accelerated repolarization[b] (hyperpolarization)
Ventricular myocardium	shortened refractory period (enhanced contractility)[a]	accelerated repolarization[b] (hyperpolarization)[b]

[a] Possibly or probably due to increased calcium conductance.
[b] Probably due to increased potassium conductance.
[c] Probably due to accelerated decrease in i_{K2}.
From *Katz* [206]; reprinted with permission.

Cells of the lower nodal region (NH) are capable of spontaneous impulse formation; in these cells catecholamines enhance spontaneous diastolic depolarization and impulse initiation. However, catecholamines do not appear to cause firing in the AN and N regions of the AV node.

Stimulation of beta-adrenergic receptors of ventricular muscle cells has little effect on transmembrane potentials and conduction velocity; spontaneous depolarization and automatic impulse formation are not induced but repolarization may be slightly accelerated [181].

The response to catecholamines of cells with abnormal electrical properties (due to disease, drugs or experimental intervention) may differ dramatically from that of normal myocardial cells. Sometimes catecholamines may restore electrical activity of depressed cells toward normal; at other times they may facilitate or cause unusual electrical phenomena. By sufficiently hyperpolarizing transmembrane potentials of depressed cells, abnormal automatic firing frequently ceases and normal electrical activity is restored. If cells are markedly depressed by adrenergic beta blockade,

Table X. Effects of acetylcholine on the electrophysiological properties of the heart

Region	Physiological response	Effects on action potential
SA node	slowing of pacemaker	hyperpolarization[a], slowed diastolic depolarization[a]
Atrial myocardium	shortened refractory period (depressed contractility)	acceleration of repolarization[a], reduced slow inward current
AV node	slowed conduction	reduced 'summation' of impulses in AN region (acceleration of repolarization and decreased amplitude)[b]
His-Purkinje system and ventricular myocardium	little or none	acceleration of repolarization only at extremely high concentrations[a]

[a] Probably due to increased potassium conductance.
[b] Possibly due to increased potassium conductance.
From *Katz* [206]; reprinted with permission.

catecholamines may increase resting potential or maximum diastolic potential and restore transmembrane potential toward normal. The effect of catecholamines on rate, rhythm and conduction of abnormal cells altered by drugs or disease is difficult to predict; sometimes catecholamines restore electrical activity toward normal but at other times they may cause after-potentials and 'slow' responses (i.e. action potentials with remarkably low rates of depolarization, which are small in amplitude and propagate extremely slowly) [181].

The effects of vagal stimulation or acetylcholine administration on electrical activity of normal and abnormal myocardial cells will not be considered here. Recent discussions of this topic appear elsewhere [181, 393]. However, for simplicity of comparing and understanding differences between the effects of catecholamines and acetylcholine on the physiological response and action potential in various regions of the heart, tables IX and X have been included.

Normally, changes in cardiac activity caused by the sympathetic or parasympathetic nerves result from increased activity in one of these components of the autonomic system and from decreased activity in the other. Sympathetic stimulation increases sinus rate and, since AV conduction is enhanced and the sinus normally fires at the most rapid rate, sympathetic stimulation does not cause arrhythmias despite vagal withdrawal. However, failure of the sinus to respond to the liberated norepinephrine or impairment of impulse transmission from sinus to atrium may be accompanied by emergence of ectopic atrial or ventricular rhythm; under these circumstances, sympathetic stimulation increases normal automaticity of specialized cells at ectopic sites. Even if the sinus responds normally to the liberated norepinephrine, enhanced responsiveness to catecholamines elsewhere in the heart can lead to arrhythmias and conduction disturbances. If AV conduction does not permit a one-to-one transmission of a rapid atrial rate, ventricular escape may occur.

Abnormal cells in the conduction system may respond to normal concentrations of norepinephrine with an excessive rate and escape from sinus control. Activation of some efferent sympathetic nerves may change conditions in only one area of the heart and alter impulse conduction, impulse generation or repolarization.

Changes in electrophysiologic properties of cardiac muscle which occur in various diseases and during therapy with some drugs may markedly influence the response of the heart to sympathetic activation. For example, in the presence of a segment of ischemic myocardium, the inotropic and chronotropic effects caused by augmented sympathetic tone will increase perfusion requirements and intensify ischemia. The latter may then evoke arrhythmias and conduction disturbances. Electrophysiologic alterations occur with some drugs; digitalis may modify the response of myocardial fibers to catecholamines and thereby induce arrhythmias during sympathetic nerve activation.

Sympathetic activity may also improve AV conduction, reduce heart block, eliminate an ectopic ventricular focus and restore cardiac rhythm. Similar effects may be exerted on supraventricular rhythm in the presence of sinoatrial conduction defects. An increased sinus rate may actually reduce the likelihood of atrial and ventricular ectopic impulses.

The interplay of the autonomic nervous system and its effects on abnormal electrophysiology of the heart is complex and the myocardial responses variable and sometimes unpredictable. For a detailed account of neural influences on cardiac electrical activity and rhythm the reader is referred to the review by *Hoffman* [181] on which the above discussion is based.

K. Autonomic Nervous System Influence on
the Electrocardiogram and Cardiac Rhythm

Electrocardiographic wave form abnormalities may occur in patients with central nervous system disease (e.g. trauma, tumor, infections, cerebrovascular accidents). Such abnormalities appear to be caused by altered autonomic influence on ventricular recovery and cardiac rhythm [2]; they consist of prolonged QT intervals, prominent U waves, and large T waves (of either polarity). Low T waves and/or ST segment displacement may also be observed but are less marked and less characteristic. Of major significance is the enhanced susceptibility of the heart to cardiac arrhythmias which occur in some patients with central nervous system disease.

Sympathetic nerve stimulation can alter T wave amplitude and cause ST segment displacement, and stellate ganglion stimulation can cause electrocardiographic and vectorcardiographic changes [334, 390, 436]. Furthermore, right ganglionectomy can cause a greater T wave positivity and a prolonged refractory period of the anterior ventricular walls, whereas left ganglionectomy can cause a greater T wave negativity and prolonged refractory periods principally in the posterior wall [436]. From the electrophysiologic effects of stimulating various cardiac nerves, it appears that there is localized rather than diffuse cardiac distribution of sympathetic nerves [2].

Central nervous system disease is not only associated with electrophysiologic abnormalities in waveforms but also with ventricular arrhythmias. Experimental procedures (e.g. stimulation of the hypothalamus or stellate ganglia) result in both waveform abnormalities and arrhythmias such as sinus and junctional tachycardia and ventricular arrhythmias. The fact that sympathetic nerve blockade can sometimes eliminate these abnormalities supports the view that autonomic dysfunction is the cause of the abnormalities.

Prolongation of the QT interval can be induced by left stellate ganglion stimulation and by right stellate ganglionectomy. Also, it has been reported that the rapid intravenous injection of catecholamines as a bolus caused transient QT prolongation, whereas slow infusion shortened the QT interval. It was suggested that administration of catecholamines in a bolus may not permit complete intravascular mixing; therefore, the catecholamine effect on different parts of the heart may have been unequal. Prolonged QT interval syndromes have been described in man and in this condition, ventricular fibrillation, particularly after exertion or emotional stress, may result in syncope. This condition has been successfully treated by left stellate ganglionectomy and left sympathectomy, and propranolol has also been employed successfully to diminish episodes of ventricular fibrillation. Evidence suggests that prolonged QT interval syndromes involve 'a local imbalance of cardiac responses to the sympathetic nervous system' [2].

Experimental sympathetic stimulation reduces threshold and increases vulnerability of the heart to ventricular fibrillation [217]. If both sympathetic stimulation and coronary occlusion are induced simultaneously, then the fibrillation threshold is reduced even further. On the other hand, sympathectomy raises fibrillation threshold and protects against fibrillation. Vagal stimulation also elevates fibrillation threshold and decreases the occurrence of spontaneous fibrillation induced by coronary occlusion. Excess parasympathetic activity can account for the reflex bradycardia and atrioventricular conduction disorders associated with some types of myocardial infarction [2].

Increased adrenergic activity may accompany myocardial infarction as a result of fear, reduced cardiac output and decreased blood pressure; evidence of increased adrenergic activity is reflected by increased plasma and urinary catecholamine concentrations [139, 179, 199, 274, 295]. The effectiveness of adrenergic beta blockade in reducing the occurrence of arrhythmias further supports the concept that adrenergic activity plays an important role in the genesis of arrhythmias in ischemic heart disease and myocardial infarction. Whether the antiarrhythmic action of propranolol was due primarily to its beta-adrenergic blocking effect or its quinidine-like effect on membrane stabilization was uncertain [235]. However, since cardioselective beta blockers (e.g. practolol and satalol), which have minimal quinidine-like actions, are effective in supressing some types of arrhythmias due to ischemic heart disease, it seems probable that their antiarrhythmic action results from beta blockade [215]. The arrhythmogenic threshold of the myocardium may be reduced by either the norepinephrine released as the neurotransmitter from the sympathetic nerves directly innervating the myocardium or the circulating catecholamines released from the adrenal medulla or sympathetic nerves elsewhere in the body. It is noteworthy that practolol has been reported to have adrenergic nerve depressant properties [327] and that it is more effective in treating arrhythmias occurring early when plasma catecholamine concentrations are usually highest rather than late after myocardial infarction [5]. Some studies on digitalis and ouabain-induced toxicity suggest that practolol exerts its antiarrhythmic effect not only by beta-adrenergic receptor blockade but by depressing adrenergic nerve activity [211].

Finally, it has been pointed out that increased sympathetic activity to the heart can increase the magnitude of ischemic injury in the experimental animal and thereby enhance the genesis of arrhythmias [2]. The degree of ischemic injury and the occurrence of arrhythmias may be reduced by propranolol which has been shown to improve myocardial oxygenation and hemodynamics in some patients following myocardial infarction [288].

V. Adrenergic Involvement in Cardiac Pathophysiology

From the foregoing, it seems well established experimentally that the sympathetic innervation of the heart and circulating catecholamines can play an important role in altering the electrophysiology of the myocardium and in the genesis of arrhythmias. Although less clearly defined, it appears that in the human, the adrenergic system can also play a significant or key role in some types of cardiac pathophysiology.

A. Autonomic Denervation and Arrhythmias in Human Cardiac Transplants

Recently *Harrison and Mason* [170] reported studies on human cardiac transplantation which provides a unique opportunity to assess the effect of catecholamines and adrenergic innervation on cardiac conduction and arrhythmias. Their studies are particularly pertinent since the central nervous system, via the autonomic system, may play a role in altering cardiac electrophysiology and in generating arrhythmias and in increasing vulnerability to ventricular fibrillation [140, 246, 247]. There is also evidence that elevated levels of catecholamines can cause arrhythmias, particularly in the presence of ischemic areas in the heart [282, 336]. Furthermore, clinical studies have demonstrated that the use of beta-adrenergic blockers (e.g. practolol or alprenolol) reduce the incidence of sudden death in the late postmyocardial infarction period [289, 420].

Arrhythmias occur in the majority of cardiac transplant patients (table XI). Atrial arrhythmias are often a sign of early cardiac rejection. From the experience of *Harrison and Mason* [170] it appears that sympathetic innervation of the heart is not required for the genesis of ischemic or non-ischemic ventricular arrhythmias. They also found that in these human cardiac allografts the conduction system is responsive to administration of catecholamines and adrenergic blocking drugs in a manner similar to the

Table XI. Arrhythmias in cardiac transplant patients

Type of arrhythmia	Total number of patients	Total number of episodes
Atrial	47	159
Atrial premature beats	26	110
Atrial bigeminy	2	22
Atrial flutter	4	10
Paroxysmal atrial tachycardia	7	8
Atrial fibrillation	6	7
Sinus arrest	2	3
Junctional rhythms	18	33
Ventricular	27	88
Ventricular premature beats	24	83
Ventricular fibrillation	5	5

From *Mason* et al. [269]; reprinted with permission.

normal heart – suggesting a normal adrenergic receptor mechanism post-transplantation. As mentioned previously, despite a decreased rate of response to exercise, the physiologic function of the heart appears normal and permits normal activity for these patients.

Finally, *Harrison and Mason* [170] reported that the electrophysiologic studies of the sinus node, atrium and A-V node demonstrated only minor changes, due probably to lack of autonomic innervation. The electrophysiologic action of drugs on the denervated human heart are indicated in table XII. More detailed account of the effects of catecholamines and adrenergic innervation on cardiac conduction and arrhythmias in the human may be found elsewhere [170].

B. Arrhythmias in Myocardial Ischemia and Infarction

Earlier in this monograph studies were cited which indicate that an augmented adrenergic activity to the heart facilitates the induction of ventricular arrhythmias including fibrillation. Ventricular tachycardia is frequently induced by electrical stimulation of sympathetic nerves to the heart [8, 141, 163, 392, 396]. Electrophysiological evidence indicates that coronary occlusion can activate cardiac sympathetic afferent fibers [252], which in turn can excite efferent cardiac sympathetics [253] and result in ventricular

Table XII. Electrophysiologic actions of drugs[a] in the denervated human heart

Drug	Auto-maticity	Sinus node recovery	A–V node refrac-toriness	A–V node conduc-tion velocity	His-Purkinje velocity	Comment
Norepinephrine	↑	?	↓	↑	0	no definite super-sensitivity
Isoproterenol	↑	shortens	↓	↑	0	similar to inner-vated heart
Propranolol	↓	prolongs?	↑	↓	0	receptors intact
Atropine	0	0	?	?	0	demonstrated lack of vagal innervation
Acute digoxin	?	0	0	0	0	acts by vagal stimulation
Chronic digoxin	?	?	↑	↓	0	acts directly without neural connection to ↑ Wenckebach block
Quinidine	↓	0	↑	↓	↓	acts directly and via neural mechanism
Edrophonium	0	0	0	0	0	acts only on remnant atrium of recipient

[a] Effects shown in therapeutic concentrations and doses.
↑=Increases; ↓=decreases; 0=no effect; ?=not known.
From *Harrison and Mason* [170]; reprinted with permission.

arrhythmias [141]. Bilateral cardiac sympathectomy reduces the incidence of ventricular fibrillation and the mortality rate which usually follows experimental coronary occlusion [78, 99, 101, 125, 167, 168, 361]. Furthermore, beta-adrenergic blockade exerts a protective effect in reducing ventricular fibrillation following coronary occlusion [215] and in shortening the duration of arrhythmias [75]. In contrast to the deleterious effect of sympathetic activation on cardiac rhythm following coronary occlusion, activation of the parasympathetic nervous system appears to exert a beneficial effect on rhythm [140].

Table XIII. Myocardial norepinephrine[a]

	BSTG-4 wk	BSTG-6 mo	Control
Left atrium mean		0.89±0.28 (9)	1.0 ±0.4 (7)
Nonischemic ventricle			
Mean	0.13±0.08 (3)	0.66±0.3 (11)	0.68±0.4 (15)
<15 min		0.75±0.33 (4)	0.86±0.5 (6)
>48 h		0.51±0.25 (5)	0.59±0.33 (8)
Infarcted ventricle			
Mean		0.59±0.38 (7)	0.58±0.5 (10)
<15 min		0.69±0.37 (4)	0.94±0.63 (4)
>48 h		0.21±0.06 (2)	0.3 ±0.23 (5)

[a] Values in micrograms per gram of myocardium (wet weight ± standard deviation). BSTG = Bilateral stellate ganglionectomy and thoracic sympathectomy; 4 wk = 4 weeks duration; 6 mo = 6 months duration; <15 = animals dying in less than 15 min; >48 = animals surviving more than 48 h. Number of dogs in each group in parentheses. Mean values for BSTG-4 wk were significantly less than mean values for control and for BSTG-6 mo ($p < 0.005$, derived by Student's t-test) for the nonischemic ventricle. No other significant differences were found.
From Fowlis et al. [125]; reprinted with permission.

Ebert [99] showed that in dogs with cardiac sympathectomy, the release of potassium from the myocardium was absent or attenuated following coronary ligation. Therefore, activity of the sympathetic innervation of the heart may be an important determinant of potassium flux and may be involved in the genesis of ventricular arrhythmias accompanying a coronary occlusion. It is noteworthy that the size of infarcts resulting from coronary occlusion were similar whether the cardiac nerves were intact or not [322].

Fowlis et al. [125] demonstrated that cardiac sympathectomy provided significant protection against postocclusion mortality, particularly in the first 15 min following occlusion. Thereafter, the difference in mortality following coronary occlusion between control dogs and dogs with cardiac sympathectomy became progressively less significant with the passage of time. Although the reasons for the protective effect of sympathectomy are unknown, these investigators pointed out that removal of sympathetics may result in depletion or impaired release of myocardial norepinephrine, a diminished stimulus to the release of adrenal catecholamines into the circulation, interruption of cardio-cardiac excitatory spinal reflexes, or a combination of these mechanisms.

In accord with the findings of others [101, 154], *Fowlis* et al. [125] demonstrated that myocardial norepinephrine was significantly reduced (0.13 ± 0.08 µg/g of myocardium, wet weight) 4 weeks after cardiac sympathectomy but returned to normal (0.66 ± 0.3 µg/g) at 6 months. (The explanation for this return to normal concentration of norepinephrine is unclear. Others have reported that cardiac sympathectomy results in a marked depletion in myocardial catecholamines for as long as 12 months, despite reinnervation [162].) Slightly higher norepinephrine concentrations were found in the nonischemic myocardium of the left ventricle of dogs that succumbed within 15 min of the coronary occlusion as compared to ventricles of dogs surviving more than 48 h; the lower norepinephrine concentration in the latter group may have resulted from some degree of heart failure. Norepinephrine concentrations in infarcted areas of the ventricles gradually decreased over 48 h to relatively low levels; this finding is similar to results reported by others [337]. Table XIII reveals the effect of coronary occlusion on norepinephrine concentrations in intact dogs and those subjected to cardiac sympathectomy.

Some investigators [271] have reported that experimental myocardial infarction causes a profound decrease or complete disappearance of norepinephrine concentrations in both necrotic infarcted areas as well as the non-infarcted myocardium – more so in ventricular than atrial tissue; repletion of myocardial norepinephrine started 2 weeks following infarction and normal concentrations were found by 6 weeks.

Fowlis et al. [125] found no correlation between myocardial norepinephrine content and postocclusion heart rate and ventricular arrhythmias. Hence, they could not implicate the absolute norepinephrine concentration in the genesis of arrhythmias and death. Whether coronary occlusion in dogs with cardiac sympathectomy may be associated with impaired release of myocardial norepinephrine and/or an impaired reflex release of catecholamines into the circulation from the adrenal and elsewhere remains uncertain. Although the precise mechanism whereby sympathectomy protects against fatal arrhythmias following occlusion is unclear, a suppression of the adrenergic effect on the heart appears to be implicated.

Spurgeon et al. [372] reported that cardiac denervation resulted in a reduction of atrial and ventricular epinephrine to 40–60% of control concentrations, whereas norepinephrine was reduced to 1–3% of control levels. They also found that the norepinephrine in conductile tissue (i.e. the bundle branches, SA and AV nodal tissue) was reduced to levels approaching zero, whereas epinephrine concentrations in conductile tissue (except for the SA node) were not significantly reduced. These results suggested that, in contrast to the intraneuronal location of myocardial norepinephrine, epinephrine appears to be present in a nonneuronal location – since it persists after denervation.

Catecholamines can increase diastolic depolarization in Purkinje fibers and pacemaker cells, and thereby can induce repetitive firing [180]. As pointed out by *Randall* [322, pp. 43–94], these changes induced by catecholamines, coupled with the nonuniform recovery of ventricular muscle excitability [166], can result in disorganized conduction and lead to multiple reentry circuits and fibrillation. *Randall* [322, pp. 43–94] has stated that: 'The high incidence of premature beats and ventricular tachycardia in normal dogs subjected to major coronary vessels occlusion and the marked attenuation of such dysrhythmias following denervation suggest an important association between local catecholamine content of the myocardium and the development of arrhythmias after infarction. A homogenous distribution of catecholamines throughout the myocardium, on the other hand, tends to decrease vulnerability to such dysrhythmia' [165].

Opie et al. [298] have recently reviewed the biochemical aspects of arrhythmogenesis and ventricular fibrillation. They indicated that any biochemical event that (a) decreases resting potential or (b) produces the slow response or (c) shortens the action potential duration or (d) enhances phase 4 depolarization could be important in the genesis of arrhythmias. They proposed that acute myocardial ischemia is accompanied by an abrupt increase in extracellular potassium (within seconds of coronary occlusion) which causes a loss of membrane potential and a shortened action potential duration. Furthermore, localized hyperkalemia can block the fast response (i.e. the fast inward current caused by sodium ions passing through inward channels in myocardial cells) and unmask or evoke a slow response (i.e. the slow inward current caused predominantly by calcium ions passing through slow inward channels). If hyperkalemia is sufficiently severe it can prevent normal depolarization, and under such circumstances a slow response may be provoked by catecholamines via the generation of cyclic AMP in myocardial cells. In experimental coronary occlusion, increased cyclic AMP has been linked to ventricular arrhythmias and fibrillation. This hypothesis is supported by experiments in which perfusion of an isolated heart preparation with cyclic AMP, or beta-adrenergic agonists, or theophylline caused a decreased ventricular fibrillation threshold. Also, increased automaticity occurs in partially depolarized Purkinje fibers, especially when there is stimulation by catecholamines or cyclic AMP in the presence of a low external potassium level [298]. It is noteworthy that at supposed anti-arrhythmic concentration of propanolol, the following effects are exerted on Purkunje fibers: (a) decreased action potential amplitude; (b) decreased maximal slope of phase-0 depolarization; (c) decreased membrane responsiveness; (d) decreased conduction velocity; (e) decreased action potential duration; and (f) decreased automaticity [182]. Propranolol also slows sinus automaticity and increases the effective refractory period of atrial and AV nodal tissue and thus antagonizes the electrophysiologic effects of beta-adrenergic stimulation. The P-R interval is prolonged by propranolol's action on the AV node (increasing the A-H but not the H-V interval). A high degree of AV nodal block by propranolol must be avoided clinically since subsidiary pacemakers may be suppressed and asystole result [10]. It should be appreciated that in addition to its action as a beta-adrenergic blocker, propranolol (at relatively high concentrations) also exerts a direct membrane effect somewhat similar to that of quinidine (e.g. propranolol inhibits sodium entry and increases potassium flux and potassium content in myocardial cells [10]).

It is noteworthy that the regional myocardial dysfunction and hemodynamic abnormalities observed during strenuous exercise in dogs with limited coronary flow (i.e. partial constriction of the circumflex coronary artery) [387] can be significantly improved by the administration of propranolol [388]. Heart rate decreased, contraction of the ischemic myocardial segment increased markedly, as did left ventricular wall thickening, and coronary flow returned to a normal velocity pattern; these favorable effects were only partially diminished by pacing the heart to increase heart rate. It was concluded that the main beneficial effect of propranolol was to reduce heart rate and thereby decrease the myocardial oxygen requirements imposed by the tachycardia that invariably accompanied coronary constriction. However, the results of pacing suggested that factors other than the effect of propranolol on the heart rate played a role in its beneficial effect [388].

C. Myocardial Ischemia and Infarction

Maroko, Braunwald and associates have presented convincing evidence which indicates that a variety of metabolic and pharmacologic interventions can alter myocardial infarct size following coronary occlusion [31, 261]. Their findings support the concept that the extent of ischemic myocardial damage following coronary occlusion is importantly influenced by the balance between myocardial oxygen supply and demand. Furthermore, it appears that myocardial damage can be reduced or minimized by interventions which (1) reduce oxygen demand, (2) increase myocardial oxygen supply, (3) augment anaerobic metabolism, (4) enhance transport of energy substrators, or (5) protect against autolytic or heterolytic damage [261].

Maroko and associates [263] also demonstrated that pacing-induced tachycardia or administration of agents causing a positive inotropic or chronotropic effect on the heart (e.g. isoproterenol, glucagon, bretylium, digitalis) will elevate myocardial oxygen consumption and will increase the extent of myocardial injury following experimental coronary occlusion. Therefore, it is conceivable that in some patients following myocardial infarction a significant activation of the adrenergic system may increase the extent of myocardial injury. According to a recent report, based on serial determinations of plasma catecholamine concentrations in patients with acute myocardial infarction, it appeared that the magnitude of adrenergic activation was related to the extent of myocardial damage and late mortality [205].

Contrariwise, reducing oxygen consumption by beta-adrenergic blockade with propranolol [261–263] or practolol [241] reduced myocardial ischemic injury following experimental coronary occlusion. One recent experimental study on rats revealed that a combination of alpha and beta blockade with labetalol was more effective than only a beta blockade with propranolol in preventing myocardial necrosis after acute coronary occlusion [57].

Although use of beta blockers (propranolol or metoprolol) within 4–7 h following myocardial infarction has been reported to decrease infarct size in man, differences in mortality and morbidity could not be detected [286]. *Mueller* [286] found that this therapeutic effect of propranolol (administered 5–13 h after infarction) was obvious in larger infarcts but not smaller ones – perhaps due to their technique of assessing infarct size but possibly due to the fact that larger infarcts cause greater adrenergic stimulation and thus are more sensitive to beta-adrenergic blockade.

In patients with myocardial infarction adequate doses of propranolol can effectively reduce mean arterial pressure heart rate, cardiac work and myocardial oxygen consumption and can also increase myocardial extraction of lactate from arterial blood; improvement in the balance between myocardial oxygen supply and demand can apparently be safely accomplished by beta-adrenergic blockade [288].

As stated by *Singh and Burnam* [359]: 'Beta-adrenergic blocking drugs have the potential for modifying in a beneficial manner the ischemic consequences of coronary artery stenosis. These compounds influence the major determinants of myocardial oxygen supply and demand and may reverse the arrhythmogenic propensity and oxygen-wasting effect of catecholamines, the plasma and urine levels of which are increased during early phases of myocardial infaction.' They have emphasized that adrenergic blockade can be of value in controlling supraventricular arrhythmias complicating acute myocardial infarction and in the treatment of resistant or recurrent ventricular arrhythmias, particularly if they are related to augmented adrenergic activity.

It is of considerable interest that a recent multicenter study demonstrated rather conclusively that when long-term treatment with timolol (a noncardioselective beta-adrenergic blocking agent) was started 7–28 days after myocardial infarction both mortality and reinfarction rates were significantly reduced by 44.6 and 28.4%, respectively [297]. Although the beneficial effect of timolol is probably related mainly to reduction of cardiac oxygen demand, *Sleight* [362] has pointed out that it may also result from blocking the arrhythmogenic effects of excess catecholamines or by reducing free fatty acids which may also be arrhythmogenic and may result from excess adrenergic activity. Others have also reported studies suggesting that in patients who have recovered from myocardial infarction the occurrence of sudden death was markedly reduced by treatment with beta-adrenergic blocking drugs [3, 156, 420].

Intra-aortic balloon counterpulsation reduces myocardial oxygen needs by lowering systolic pressure and augmenting coronary artery flow (by increasing aortic diastolic pressure); this mechanical intervention was found to markedly reduce the extent of myocardial ischemic injury [35, 260]. In addition *Spotnitz* et al. [371] have observed that intra-aortic balloon counter pulsation appears to reduce adrenergic activity and minimize release of catecholamines into arterial and coronary sinus blood; conceivably this limitation of adrenergic activity may play an additional role in reducing ischemic injury by intra-aortic balloon counterpulsation.

It is pertinent that in some patients who develop S-T segment elevation (as a consequence of arterial hypotension, anginal pain, or ventricular fibrillation) a reduction in the S-T elevation may occur after propranolol administration [260, 264] and during intra-aortic balloon counterpulsation [260]. In addition, propranolol administration [144] and intra-aortic balloon counterpulsation [234] reduced the S-T segment elevations and improved metabolism in patients with acute myocardial infarction [287, 288].

Despite the potentially harmful effects which may be caused by epinephrine and norepinephrine on an ischemic area of myocardium, dopamine (Intropin) has been employed therapeutically in the treatment of cardiogenic shock resulting from myocardial infarction. Dopamine exerts an inotropic effect on the heart, usually without a significant increase in heart rate, and thereby causes a rise in systolic blood pressure and improves peripheral blood flow. Dopamine causes vasoconstriction of skeletal muscle vasculature; however, the unique ability of this catecholamine to dilate renal, mesenteric, cerebral and coronary arteries (as mentioned in section III) are distinct virtues in the treatment of cardiovascular shock – particularly when the blood pressure is not at a critically low level which may initially require a potent vasoconstrictor.

The treatment of cardrogenic shock or congestive heart failure accompanying myocardial infarction remains a controversial and difficult therapeutic problem. *Goldstein* et al. [151] have recommended that in patients with acute myocardial infarction and heart failure the synthetic catecholamine Dobutamine (Dobutrex) be used before digoxin and other catecholamines. They have further suggested that Dobutamine is preferable to large doses of diuretics, since diuretics may compromise cardiac function in patients with myocardial ischemia who require a high ventricular filling pressure to maintain cardiac output. The virtues of this drug are that it has a potent positive inotropic action with little chronotropic arrhythmogenic effects and also a short half-life (2.5 min).

Goldstein et al. [151] reported that, in patients with myocardial infarction, dobutamine increased cardiac output (by 33%) and stroke work index, and decreased left ventricular filling pressure and systemic vascular resistance without changing heart rate or arterial blood pressure; relief of pulmonary congestion resulted and both preload (pulmonary arterial occlusive pressure) and afterload (systemic vascular resistance) decreased. As they pointed out, the decrease in intraventricular pressure without a significant change in perfusion pressure would be expected to increase coronary perfusion. In contrast, they found that digoxin administered intravenously had no effect on cardiac index and only slightly increased stroke work index and did not affect preload or afterload. Furthermore, they cited the finding that digoxin can enhance myocardial ischemia [151].

As mentioned by *Goldstein* et al. [151], isoproterenol causes hypotention and tachycardia, decreases coronary flow, increases oxygen demand and markedly enhances ventricular arrhythmias and myocardial ischemia – effects similar to those caused by norepinephrine. Finally, they pointed out that dopamine is less than ideal in the treatment of cardiac failure accompanying myocardial ischemia. Although it does exert a positive inotropic effect similar to that of dobutamine, dopamine increases heart rate substantially and in high dosage it increases peripheral resistance and may thus markedly increase left ventricular filling pressure.

It is particularly intriguing and significant that a recent study of patients with 'preinfarction' angina documented a vasospastic origin of their anginal attacks [266]. *Maseri* et al. [266] observed a reduction in myocardial blood supply and the presence of coronary vasospasm during angina, independent of the severity of coronary artherosclerosis. They further demonstrated that angina at rest with ST-segment depression was not preceded by increased myocardial metabolic demand but was associated with reduced myocardial perfusion and coronary vasospasm. It is also noteworthy that other investigators have demonstrated that coronary vasoconstriction can be reflexly induced in patients with effort angina by the cold pressor test [285].

Maseri et al. [266] emphasized that coronary artery spasm has been established as the cause of Prinzmetal's angina and that this mechanism may be extended to include a significant number of other forms of angina. They concluded that coronary vasospasm may be an important cause of myocardial infarction and thrombosis, in the absence or presence of a variable degree of coronary atherosclerosis.

Very recently, using quantitative angiography and alpha-adrenergic stimulation in patients with spontaneous rest angina, *Brown* [42] observed that vasomotor hyperreactivity was localized only to the region of a pre-existing coronary atheroma. He emphasized that patients with angina, myocardial infarction, and sudden death almost always have coronary atherosclerosis which may be linked to the coronary vasospasm which plays a role in these ischemic syndromes.

Hillis and Braunwald [178] reviewed the current understanding of coronary artery spasm. They pointed out that in anesthetized animals coronary blood flow is regulated primarily by the metabolic demands of the heart and that neural influences are relatively unimportant in the control of myocardial blood flow.

On the other hand, experiments in conscious animals have demonstrated that coronary blood flow is influenced not only by metabolic requirements of the heart but can also be greatly influenced by neural mechanisms. As mentioned earlier, both alpha- and beta-adrenergic receptors have been identified in the coronary arteries of the unanesthetized dog [311]. The initial large segments of the coronary arteries are richly supplied with alpha-adrenergic receptors but are termed 'conductance' vessels since vascular resistance is mainly regulated by changes in the lumen of the smaller coronary branches which have been termed 'resistance' vessels [36]. Under normal circumstances the large 'conductance' coronary arteries contribute very little to vascular resistance; however, with adrenergic stimulation these arteries constrict markedly and increase vascular resistance [178]. There seems to be no question that in normal conscious animals and human beings neurogenic influences on the heart are potentially of great importance.

Catheter-induced coronary artery spasm has been demonstrated to occur occasionally during coronary arteriography, presumably due to mechanical stimulation of the coronary ostia. The coronary artery spasm which causes Prinzmetal's variant angina is usually accompanied by pain at rest (which usually occurs at about the same time each day) and is usually accompanied by ST elevations (indicating transmural ischemia) rather than the ST segment depressions (indicating subendocardial ischemia) occurring with typical angina. *Hillis and Braunwald* [178] cautioned that although beta-adrenergic blockade is usually beneficial to patients with typical angina pectoris, it may be detrimental in patients with Prinzmetal's variant angina since such a blockade permits alpha-receptor-mediated coronary vasoconstriction to occur unopposed. On the other hand, alpha-adrenergic blockade or beta-adrenergic stimulation may be beneficial [178]. It is noteworthy that partial cardiac sympathectomy in association with coronary artery grafting has proved beneficial to 2 patients with Prinzmetal's angina [158].

The precise cause and mechanism for the coronary artery spasm in Prinzmetal's syndrome remains unclear. Some attention has been focused on alpha-adrenergic response in the large coronary arteries resulting from sympathetic nerve stimulation or circulating catecholamines [437]. Some evidence suggests that increased sympathetic activity may result from increased parasympathetic activity; however, any interrelationship between the parasympathetic nervous system, the adrenergic nervous system and circulating catecholamines remains to be elucidated [186, p. 1194].

Hillis and Braunwald [178] have briefly reviewed the accumulating evidence that coronary artery spasm may be a cause of typical angina pectoris and myocardial infarction. Although the frequency with which coronary spasm plays a role in the genesis of typical angina pectoris and myocardial infarction is unknown, these investigators have indicated the need to assess the efficacy of alpha-adrenergic blockade in alleviating the occurrence of angina and myocardial infarction. It is interesting that alpha-adrenergic blockade with phenoxybenzamine can prevent the reflex coronary vasoconstriction elicited in patients with coronary artery disease [285]. *Braunwald* [32] made the intriguing speculation that circulating catecholamines may stimulate the alpha-adrenergic receptors of human platelets and thereby enhance their ability to produce and release thromboxane A_2, a potent vaso-constrictor agent. It is conceivable that alpha-adrenergic blockade can also prevent release of thromboxane A_2 from platelets. *Braunwald* [32] has further emphasized that in patients in whom ischemia is due to coronary spasm with impaired oxygen delivery (rather than to augmented myocardial oxygen needs) blockade of $beta_2$ alpha-adrenergic receptors may be detrimental. Propranolol produces a nonselective beta-adrenergic blockade of both $beta_1$ receptors in the myocardium (which reduces oxygen need) and $beta_2$ receptors in the coronary arteries. Since $beta_2$-receptor stimulation induces coronary vasodilatation, blockade of this receptor will remove opposition to coronary constrictor influences [32].

Hillis and Braunwald [178] have pointed out the possibility that coronary artery denervation and interruption of adrenergic vasoconstrictor fibers may play a role in the relief of angina in patients who have undergone surgical revascularization, particularly if relief persists despite occlusion of the graft.

Considerable interest has recently focused on the use of verapamil and nifedipine (calcium antagonists – i.e. inhibitors of slow channel Ca^{++} transport) in the treatment of supraventricular tachycardias, idiopathic hypertrophic subaortic stenosis, angina, and in their ability to protect the ischemic myocardium [105, 275, 294, 422]. These drugs inhibit the contractile activity of cardiac and smooth muscle and decrease myocardial oxygen uptake. In addition, they cause coronary vasodilatation and reduce peripheral resistance. Although verapamil and nifedipine are not adrenergic blocking drugs, it has been suggested that the ability of these drugs and propranolol to protect the ischemic myocardium may involve a common mechanism – an ability to slow the rate of depletion of ATP reserves [294].

Pertinent to the therapeutic use of verapamil is the potential effects of catecholamine and drug interaction. As recently pointed out, since beta-adrenergic agonists (e.g. epinephrine and isoproterenol) or methylxanthines (e.g. caffeine and theophylline) increase cyclic AMP levels, which in turn can increase the number of functioning calcium channels, these drugs may oppose the calcium channel-blocking effects of verapamil; contrariwise, beta-adrenergic antagonists may enhance the effect of verapamil [275].

D. Adrenergic Nervous System Alterations in Heart Failure

The adrenergic system provides a means for rapid cardiovascular circulatory adjustments – i.e. changes in myocardial contractility, heart rate

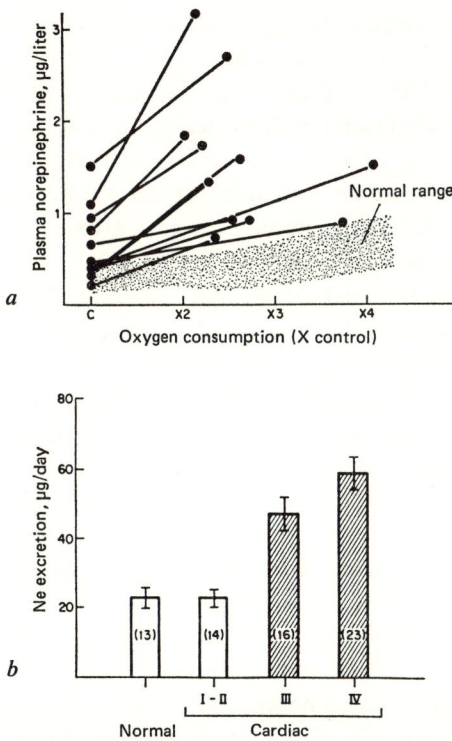

a

b

Fig. 21. Measurements of concentration of norepinephrine in patients with congestive heart failure and in normal control subjects. *a* Changes in plasma norepinephrine during exercise in congestive heart failure. Oxygen consumption during the exercise period is expressed in multiples of the resting oxygen consumption. C=Control or resting values. The normal range is represented by the stippled area. *b* Urinary NE excretion in normal control subjects, in cardiac patients without failure (classes I and II, New York Heart Association Classification) and in patients with failure (classes III and IV). The average values and their standard errors are shown. From *Braunwald* et al. [36]; reprinted with permission.

and vascular tone. Under normal circumstances activation of cardiac adrenergic nerves and/or an increase in circulating catecholamines subserve a function of 'boosting' (augmenting) cardiac function. The adrenergic system also plays an important compensatory (supportive) role in circulatory adjustments of patients to congestive heart failure. Therefore, caution must be exercised in using antiadrenergic drugs in patients with a limited cardiac reserve [37, p. 1177]. The concentration of plasma catecholamines has been used as an index of adrenergic activity; however, it must be appreciated that plasma catecholamine elevations after adrenergic activation reflect not only the magnitude of sympathetic stimulation and catecholamine release, but also the effectiveness of inactivating mechanisms and the extent of catecholamine overflow into the circulation [256]. Furthermore, it is noteworthy that physical training can markedly decrease the rate of catecholamine turnover and may, therefore, influence the adrenergic response to stress. *Östman* et al. [299, 300] have reported that cardiac norepinephrine and adrenal catecholamines are significantly increased in trained rats. With acute prolonged exercise cardiac norepinephrine decreased in untrained but not in trained rats. Cardiac norepinephrine turnover was markedly slower in trained than untrained rats. During exercise, norepinephrine turnover increased but remained slower in trained rats, and the urinary excretion of catecholamines was greater in the untrained animals. They concluded that chronic physical training induces a functional adaptation of the adrenergic system which results in a better transmitter economy during exercise. The bradycardia of the 'athlete's heart' may be due to a decreased cardiac sympathetic nerve activity caused by training [300].

In normal persons, very little change in the concentration of arterial plasma catecholamines may occur during moderate muscular exercise. *Chidsey* et al. [60] found that average plasma norepinephrine concentrations rose from 0.28 to 0.46 μg/l. These results were similar to those found in patients with heart disease but without heart failure. On the other hand, in patients with congestive heart failure, the average plasma norepinephrine concentration was 0.63 μg/l at rest and significantly increased to 1.73 μg/l during exercise. Figure 21a reveals the greater increments in plasma norepinephrine concentrations during exercise in patients with congestive heart failure than in normal subjects. Since no consistant change in plasma concentrations of epinephrine occurred in patients with heart failure, it appears that the adrenal medulla did not participate in the increased adrenergic activity. *Chidsey* et al. [61] concluded that increased adrenergic activity may have an important supportive role in patients with congestive heart failure. Elevated urinary excretion of norepinephrine provides additional evidence of augmented sympathetic nerve activity in patients with congestive heart failure (fig. 21b) [58]. Recently it has been reported that plasma norepinephrine elevations in patients with heart failure are correlated directly with the degree of left ventricular dysfunction [386]. Since it has been observed that circulating lymphocytes in patients with severe heart failure failed to generate normal amounts of cyclic AMP when exposed to isoproterenol, the question

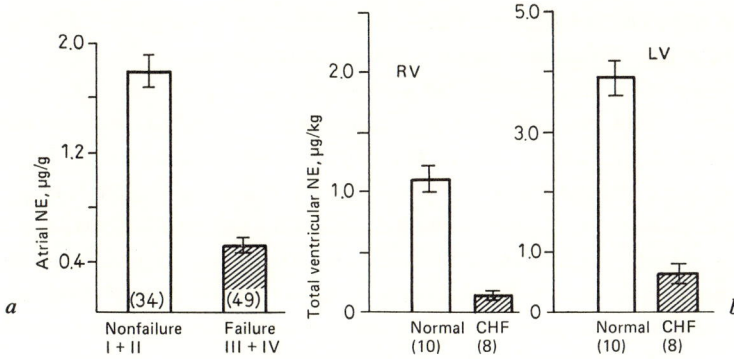

Fig. 22. Effects of heart failure on the cardiac stores of norepinephrine. *a* Concentration of NE in atrial appendage biopsies taken during cardiac operations from 34 patients without heart failure (classes I and II) and 49 patients with heart failure (classes III and IV). Average values and their standard errors are included. *b* Total ventricular NE content in normal dogs and in dogs with pulmonary stenosis, tricuspid insufficiency and congestive heart failure (CHF). Average values are given with their standard errors. RV = right ventricle, LV = left ventricle. From *Braunwald* et al. [36]; reprinted with permission.

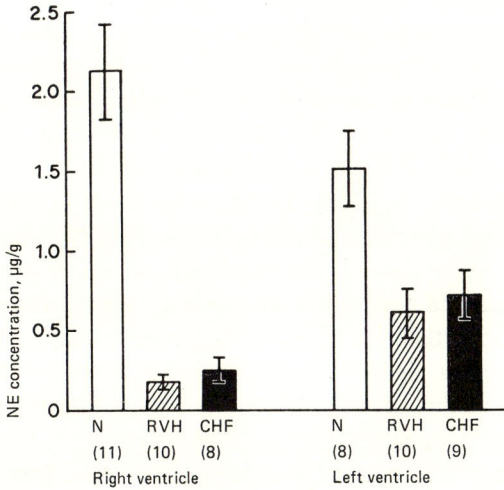

Fig. 23. The average norepinephrine concentration of the right and left ventricles of normal cats and cats with right ventricular hypertrophy (RVH) and congestive heart failure (CHF). Vertical lines with cross bars equal ±1 standard error of the mean. Numbers in parentheses equal number of animals in each group. From *Spann* et al. [367]; reprinted by permission of the American Heart Association.

has been raised as to whether a beta-adrenergic receptor desensitization may account for the dysfunction in myocardial contractility.

Chidsey et al. [58, 59, 64] also demonstrated a significant depletion of norepinephrine in myocardial tissue removed during cardiac surgery from patients with heart failure. The norepinephrine concentration in atrial tissue removed from patients with congestive heart failure averaged 0.49 µg/g whereas it averaged 1.77 µg/g in patients who had not had congestive heart failure – a highly significant difference (p< 0.01). The norepinephrine concentration in left ventricular papillary muscle of patients who had heart failure averaged 0.52 µg/g. Furthermore, there was a significant (p<0.05) correlation between norepinephrine concentration in atrial and ventricular tissue (e.g. when atrial concentrations were less than 0.40 µg/g, then ventricular concentrations averaged 0.27 µg/g; when atrial concentrations exceeded 0.40 µg/g, then ventricular concentrations averaged 0.73 µg/g (fig. 22a).

Rutenberg and Spann [338] and *Braunwald* [33] concisely reviewed the alterations of cardiac sympathetic neurotransmitter activity in congestive heart failure. Much of the following discussion is based on these reviews. *Spann, Chidsey* and co-workers demonstrated that in the dog (fig. 22b) [63] and in the cat (fig. 23) [368, 369] a profound reduction in cardiac norepinephrine occurred in experimentally induced heart failure (left or right) or in right ventricular hypertrophy with or without heart failure [366, 367]. Reduced concentrations of norepinephrine occurred in both ventricles, regardless of which ventricle was subjected to a hemodynamic strain. Total norepinephrine content was also markedly diminished reflecting a true depletion rather than a 'dilution' effect of a normal amount of norepinephrine by myocardial hypertrophy [63, 338]. The time course of changes in ventricular norepinephrine concentration and content following experimental production of congestive heart failure is indicated in figure 24. It is noteworthy that there was usually no remarkable change in the concentration of norepinephrine in tissues other than the heart during experimentally induced congestive failure in the guinea pig [369]. However, in hamsters with heart failure due to cardiomyopathy, concentrations of norepinephrine were reduced in the heart, aorta and femoral arteries [259].

The finding of an increased rate of norepinephrine efflux from the coronary sinus when left ventricular pressure was acutely increased to 100 mm Hg suggests that significant increases in ventricular pressure may be involved in the mechanism of cardiac norepinephrine depletion. Studies revealed that there was a significantly diminished uptake and/or binding of infused norepinephrine in guinea pigs with experimentally induced congestive

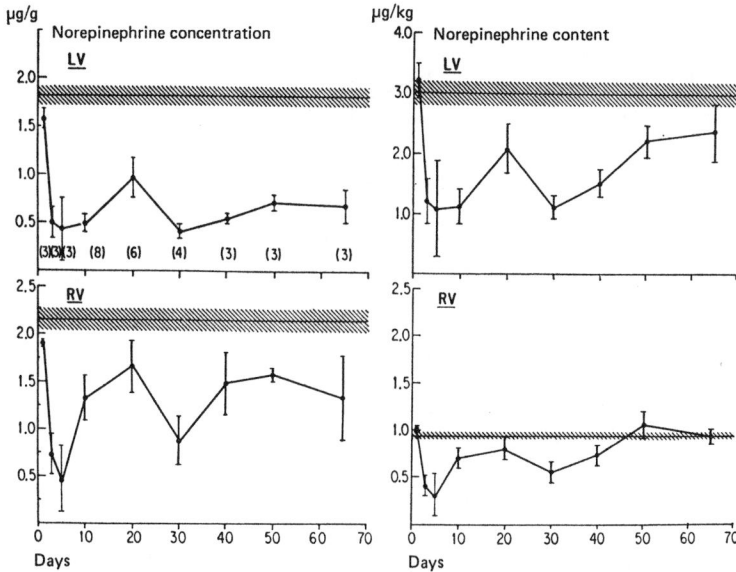

Fig. 24. Time course of changes in norepinephrine concentration in µg/g *(a, b)* and time course of changes in total norepinephrine content in each ventricle expressed as µg/kg body weight *(c, d)*. Solid circles and verticle bars represent the mean values ±1 standard error of the mean obtained from animals with congestive heart failure. Horizontal lines and hatched areas represent the mean ±1 standard error of the mean obtained from 15 normal animals. Numbers in parentheses at the bottom of *a* refer to the number of animals sacrificed at each point in time that provided the data shown in all four panels. From *Spann* et al. [369]; reprinted by permission of the American Heart Association.

heart failure as compared to control animals. This impairment in uptake or binding, or both, was evident in the heart but not the kidney of animals with congestive failure (fig. 25). Additional studies indicated that the myocardial norepinephrine depletion occurring in these animals with congestive heart failure could not be explained by a more rapid net turnover of norepinephrine. Furthermore, the depleted norepinephrine stores in the presence of a normal net turnover suggest that the rate of norepinephrine formation is reduced [369]. It has also been established that any alterations in the enzymes (COMT and MAO) responsible for inactivating the catecholamines could not account for the severe depletion of cardiac norepinephrine occurring in heart failure [221].

Pool et al. [313] found a marked reduction in cardiac tyrosine hydroxylase (the rate-limiting enzyme for norepinephrine biosynthesis) activity in dogs

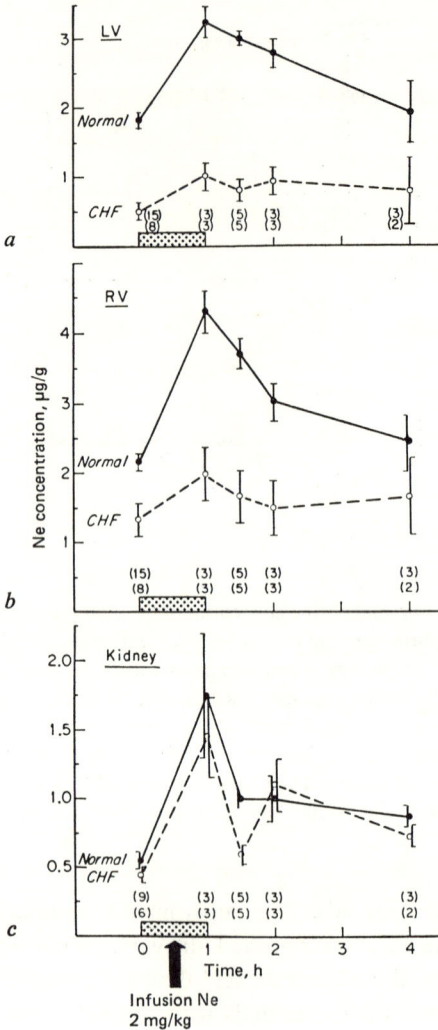

Fig. 25. Effects of infusion of norepinephrine (NE) on the concentrations of norepinephrine in the left ventricles (LV) *(a)*, right ventricles (RV) *(b)*, and kidneys *(c)* of normal guinea pigs (solid lines and circles) and guinea pigs with congestive heart failure (CHF) (open circles and broken lines). Vertical bars represent ±1 standard error of the mean. Horizontal bars represent duration of infusion, and the numbers in parentheses refer to the number of animals in each group sacrificed at the various times. From *Spann* et al. [369]; reprinted by permission.

with experimentally induced heart failure; average enzyme activity was 3.3 nmol/g/h in normal dogs compared to 0.4 nmol/g/h in dogs with congestive heart failure. Also, a highly significant positive correlation was evident between the concentration of norepinephrine and tyrosine hydroxylase in various portions of the myocardium in both the failing and normal heart. They concluded that the available evidence taken together supports the hypothesis of a parallel reduction in norepinephrine synthesis, uptake and binding of norepinephrine, rather than a specific disturbance of one of these functions [313].

The fact that reserpine depletes myocardial norepinephrine without altering tyrosine hydroxylase concentrations suggested that tyrosine hydroxylase reduction is responsible for norepinephrine depletion in cardiac failure. However, in hamsters with heart failure due to cardiomyopathy, the decrease in myocardial norepinephrine was accompanied by an *increase* in both tyrosine hydroxylase and cardiac dopamine [365]. Since immobilization of normal hamsters resulted in similar biochemical changes which reverted to normal following peripheral ganglionic blockade, it has been postulated that the increased cardiac sympathetic tone may have caused a shift in norepinephrine synthesis from hydroxylation of tyrosine to the hydroxylation of dopamine [33, 365]. It must be concluded, however, that the mechanism responsible for the defect in myocardial norepinephrine synthesis remains unknown.

Evidence indicates that in one type of experimental heart failure, there is a reduction in myocardial norepinephrine which parallels the absence of fluorescence in adrenergic nerve terminals of the heart [410]. *Vogel* et al. [410] found that in some animals recovery from heart failure was associated with a return toward normal of both norepinephrine concentration and adrenergic nerve fluorescence within 4 weeks. This interesting observation suggests that the adrenergic abnormality occurring with heart failure may be a reversible metabolic dysfunction of the neuron rather than a permanent malfunction or an actual loss of adrenergic nerves.

More recently *Coulson* et al. [76] reported that following the relief of pulmonary constriction, which was accompanied by reduction of the pressure overload hypertrophy and right ventricular failure, many indices of the hypertrophied or failing right ventricle returnerd to normal; yet, norepinephrine depletion persisted.

It is interesting that experimental Chagas' disease, produced by inoculating *Trypanosoma cruzi* into rats, caused a temporary disappearance of cardiac norepinephrine as determined by histochemical fluorescence and

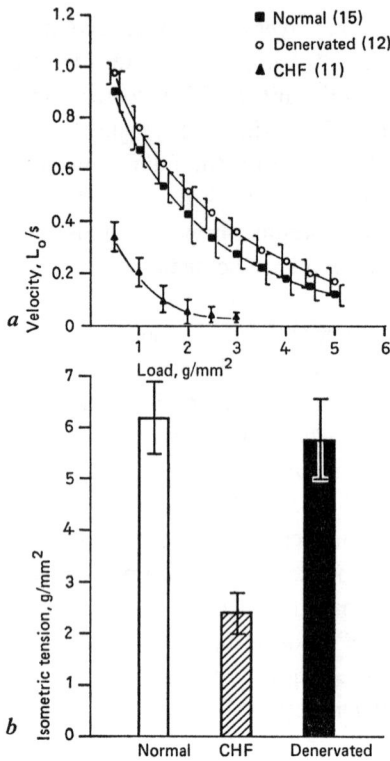

Fig. 26. Ventricular contractile state. *a* Average force-velocity in right ventricular papillary muscles isolated from normal cats and from cats with congestive heart failure (CHF) or denervation. Velocity is expressed on the ordinate as muscle lengths (L_o)/s and total load is expressed on the abscissa as g/mm^2. *b* Average maximal active tension developed at the apex of the length-tension curve in isolated right ventricular papillary muscles from the three groups of cats, expressed as g/mm^2. All vertical bars equal ± 1 standard error of the mean, and each number in parentheses represents the number of animals in that group. From *Rutenberg and Spann* [338]; reprinted with permission.

fluorometric quantitation [250]. *Machado* et al. [250] suggested that norepinephrine depletion, by impairing the positive inotropic effect of the cardiac sympathetic nerves, was probably the main factor causing congestive heart failure in the acute phase of Chagas' disease. However, it must be appreciated that undoubtedly other factors, such as the acute myocarditis, are also necessary to cause heart failure, since sympathetic denervation will not by itself produce congestive failure.

As mentioned earlier, experimental myocardial infarction is accompanied by a marked depletion of norepinephrine in the infarcted areas and also a significant decrease of norepinephrine in the noninfarcted myocardium [271, 272]. However, no correlation was noted between the concentration of myocardial norepinephrine and the first derivative of left ventricular function (dp/dt); it was concluded that alterations in left ventricular contractile function were not related to changes in norepinephrine concentrations.

Studies relating cardiac contractile function to myocardial norepinephrine concentrations were performed on ventricular papillary muscle isolated from normal cats and from those with congestive heart failure or with chronic cardiac denervation [367]. Cardiac norepinephrine concentrations were profoundly depleted in the animals with congestive failure and even more so in those with cardiac denervation. However, the contractile function of the muscle was only depressed in the cats with congestive failure; function was normal in the muscle isolated from the denervated hearts (fig. 26). Hence, it appeared that cardiac stores of norepinephrine were not fundamental for maintaining basic myocardial contractility and that norepinephrine depletion is not responsible for the intrinsic depression or myocardial contractility in the failing heart. Similar observations and conclusions have been reported by others [87].

In experimentally induced heart failure there is evidence that the failing heart is supported by circulating catecholamines arising mainly from the adrenal medulla [410]. The fact that norepinephrine-depleted heart muscle is supersensitive to the positive inotropic effect of norepinephrine [367] suggests that the circulating catecholamines may play an important role in supporting myocardial function in the failing heart of humans. This conclusion seems particularly reasonable since, as mentioned above, the circulating catecholamines are increased in patients with congestive heart failure [60]. The conclusion also gains support from evidence that congestive heart failure can appear or be aggrevated when propranolol [107, 376] or guanethidine [134] is administered to patients with advanced heart disease. In the presence of heart failure, increased circulating catecholamines may aggravate the failure by elevating vascular resistance and thereby increasing the afterload on the heart.

It is noteworthy that *Covell* et al. [77] found that despite increased adrenergic activity and the increased release of catecholamines from extra-cardiac sources which accompany congestive heart failure, the response (i.e. increments in heart rate and contractile force) to stimulation of the cardiac sympathetic nerves is markedly reduced in dogs with heart failure

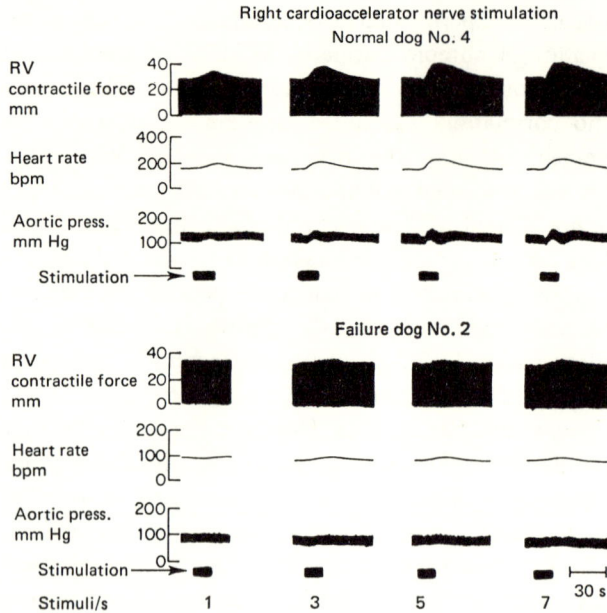

Fig. 27. Records showing the effect of right cardioaccelerator stimulation in (a) a normal dog and (b) a dog with congestive failure. From *Covell* et al. [77]; reprinted with permission of the American Heart Association.

as compared to normal animals (fig. 27). Hence, these authors concluded that the release of norepinephrine from sympathetic nerves innervating the heart must be profoundly reduced in experimental heart failure accompanied by norepinephrine depletion.

Braunwald [33] has appropriately stated that although both clinical and experimental heart failure are accompanied by marked reductions in cardiac norepinephrine and since endogenous norepinephrine does not appear to contribute to maintenance of intrinsic myocardial contractility, it appears that norepinephrine depletion occurring in congestive heart failure impairs performance not by altering intrinsic myocardial contractility but by interfering with the augmentation of contractility provided by the adrenergic system during stress. Furthermore, a diminished release of norepinephrine from cardiac sympathetic nerves in heart failure may be responsible for loss of the much needed adrenergic support of the failing heart and thus could intensify the severity of congestive heart failure.

In addition, there is evidence that the sympathetic component of the baroreflex which controls heart rate is defective in patients with heart failure [152] as well as in dogs with experimentally induced heart failure [175]. Reduction of arterial pressure with vasodilators results in an increase in heart rate and circulating catecholamines in normal subjects, whereas in patients with heart failure the increase in heart rate is blunted and the plasma catecholamines, which are already elevated, failed to increase further [69].

On the other hand, in the presence of heart failure peripheral vaso-constriction is markedly enhanced during exercise. There is evidence for a much greater reduction of mesenteric and renal blood flow caused by exercise in dogs with heart failure than in normal animals. Apparently this excessive vasoconstriction results from an increased adrenergic activity; with severe heart failure, adrenergically mediated vasoconstriction occurs in the limbs – even at rest [176, 439]. In patients with congestive heart failure and cardiac norepinephrine depletion the vasoconstrictor response to injected doses of norepinephrine was normal; yet the response to tyramine (which releases endogenous norepinephrine) was markedly increased, indi-cating that the endogenous quantity of neurotransmitter in arteriolar ventricular beds is not depleted but may actually be augmented in patients with heart failure [223].

As pointed out by *Braunwald* [33], the enhanced adrenergic vasocon-striction occurring during exercise in the presence of heart failure appears important in maintaining blood pressure and adequate perfusion of the heart and brain at the expense of other vascular beds; forthermore, intense cutaneous vasoconstriction prevents the dissipation of heat and may explain the heat intolerance seen in patients with heart failure. It is interesting that the increased cardiac output caused by cardiac glycosides is partly due to a diminution in the adrenergically mediated vasoconstriction and the regional vasodilatation which follows the administration of these glycosides [173, 268].

It should be appreciated that patients with low-output acute or chronic congestive heart failure usually have an augmented adrenergic activity which is accompanied by generalized vasoconstriction. The latter may impair cardiac output. By reducing left ventricular volume and pressure and by reducing the afterload (i.e. the impedance to ventricular ejection) vasodilators have been found beneficial in the therapy of low-output heart failure [55, 67, 68, 267, 412].

Finally, it should be mentioned that there is evidence for a reduced degree of parasympathetic influence on sinoatrial node automaticity in patients with heart failure; furthermore, the baroreceptor-induced slowing of the heart appears markedly attenuated [102]. On the other hand, in experimental heart failure, accompanied by a marked depletion of cardiac norepinephrine, a normal content of myocardial acetylcholine and a normal response to vagal stimulation has been reported [276].

E. Adrenergic Nerves and the Hypertrophied Heart

Historically it is notable that almost 30 years ago *Raab* [318] proposed that cardiac hypertrophy was induced by hormonal influences including catecholamines liberated from the adrenergic system and that hypertrophy was not simply the result of an increased work load on the heart. Subsequently, evidence has accumulated which supports the role of the adrenergic system in the development of myocardial hypertrophy.

A number of reports indicate that experimental myocardial hypertrophy is accompanied by depletion of cardiac norepinephrine concentration [84, 122, 222, 368]. However, some of these investigators have found that, although norepinephrine concentrations are reduced, the total norepinephrine content may be normal initially and only gradually decrease with time (fig. 28) [122]. Other studies reveal no significant reduction of norepinephrine concentrations in the hypertrophied hearts of spontaneously hypertensive rats [244, 433]. It is noteworthy that several investigators have reported a decrease of norepinephrine turnover in hearts of spontaneously hypertensive rats [245, 433] whereas experimental hypertension induced in the rat by administration of deoxycorticosterone and salt is accompanied by a significant increase in cardiac norepinephrine turnover [81].

Catecholamines not only regulate cardiovascular activity but they also can influence protein synthesis [434]. *Tarazi and Sen* [384] have recently discussed (a) catecholamines as inducers of cardiac hypertrophy and (b) alterations in cardiac catecholamines associated with or secondary to cardiac hypertrophy, and they have suggested some tentative implications for therapy. As they pointed out, different disturbances of catecholamine metabolism may occur in different types of hypertrophy [81, 122, 299, 429]. Most of the following is based on their discussion of findings in pressure-induced hypertension.

It is well known that experimentally isoproterenol administration can induce cardiac hypertrophy [4, 20, 331, 373, 381]. Although large doses may produce myocardial lesions and edema in addition to the hypertrophy, only hypertrophy occurs when smaller doses of isoproterenol are administered.

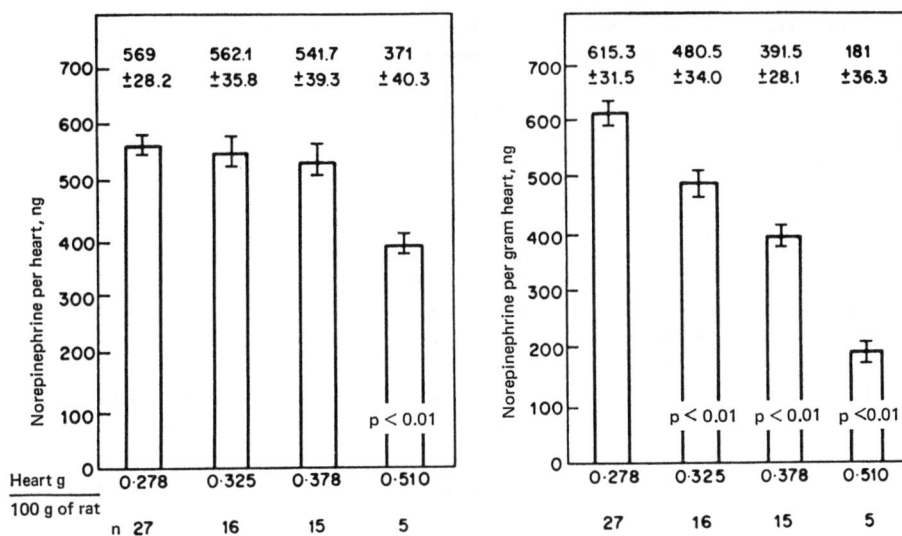

Fig. 28. Norepinephrine in hypertrophied rat hearts (constriction of abdominal aorta). With progressive hypertrophy, the concentration of cardiac catecholamine decreases, followed later by reduction of total catecholamines. From *Fischer* et al. [122]; reprinted by permission of Macmillan Journals Limited.

The increase in cardiac weight resulted from accelerated protein synthesis which was associated with an elevation of cardiac DNA and RNA [381]; furthermore, isoproterenol-induced hypertrophy could be prevented by beta-adrenergic blockade [305, 381].

Results of administering other catecholamines, even in subpressor doses, were similar to those observed with isoproterenol [137, 228], and it was postulated that the neurotransmitter, norepinephrine, might be the 'myocardial hypertrophy hormone' which, via beta-adrenergic stimulation, is responsible for the cardiac hypertrophy resulting from physical stress.

Sen et al. [352, 353] reported that two of three angiotensin antagonists [Sar[1] Ile[8]] and [Sar[1] Ala[8]] AII, caused increased myocardial protein synthesis and cardiac hypertrophy in normotensive rats. Arterial pressure was not significantly altered by these analogues of AII. They related the hypertrophy to release of catecholamines since in these hearts cardiac catecholamines were increased and the hypertrophy could be prevented by adrenalectomy prior to the administration of the analogues [353]. *Sen* and co-workers [351, 352, 384] also demonstrated that reversal of cardiac hypertrophy in the spontaneously hypertensive rat by treatment with antihypertensive drugs was

Table XIV. Antihypertensive therapy and cardiac hypertrophy in SHR

Group	BP mm Hg	Ventricular weight mg/g
Normal	120	2.6
SHR	188	3.4
Methyldopa	149	2.7
Hydralazine	123	3.4
Minoxidil	130	3.8

Results of antihypertensive therapy indicate that reversal of cardiac hypertrophy is not dependent on blood pressure control alone.
From *Tarazi and Sen* [384]. Data from *Sen* et al. [352]; reprinted with permission.

Table XV. Ventricular NE and cardiac hypertrophy response to antihypertensive therapy

Group	BP	Ventricular weight mg/g	Ventricular NE ng/g
SHR (n=18) untreated	195	3.4	520
SHR treated			
1 Methyldopa (n=8)	140[a]	2,8[a]	183[a]
2 Reserpine (n=12)	175[a]	3.1[a]	188[a]
3 'Combined'[b] (n=7)	136[a]	3.0[a]	260[a]
4 Hydralazine (n=6)	128[a]	3.5	708

[a] Statistically significant.
[b] 'Combined': mixture of reserpine, hydrochlorothiazide and hydralazine. The reversal of left ventricular hypertrophy was not dependent on blood pressure control alone; only those regimens which lowered cardiac catecholamine concentration as they reduced blood pressure led to significant reduction in ventricular weight (groups 1, 2, 3).
From *Tarazi and Sen* [384]; reprinted with permission.

not solely due to reduction of blood pressure but also depended on the ability of the antihypertensive drug to interfere with adrenergic mechanisms.

Tables XIV and XV and figure 29 reveal the effect of various antihypertensive drugs on ventricular weight and cardiac norepinephrine concentration. The ability of antihypertensive drugs to reverse cardiac hypertrophy was correlated with their ability to reduce cardiac norepinephrine [384].

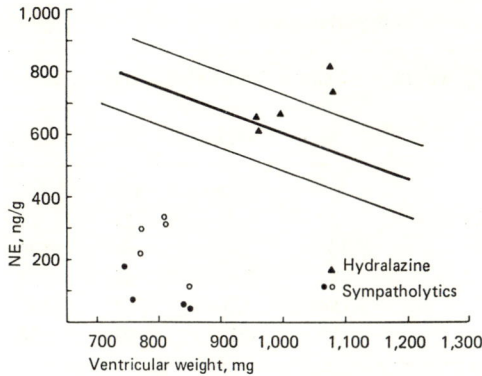

Fig. 29. Concentration of norepinephrine (NE) in myocardium was found to be inversely related to ventricular weight in both normal (WKY) and weight-matched spontaneously hypertensive rats (SHR; regression line and confidence limits of equation are shown in the graph, see also fig. 30). Against this background are plotted the results of treatment of SHR by reserpine or methyldopa (sympatholytics) and by hydralazine; all three drugs lowered arterial pressure but reversal of cardiac hypertrophy was associated with concomitant reduction of ventricular NE. From *Tarazi and Sen* [384]; reprinted with permission.

Prolonged beta-adrenergic blockade with propranolol has been reported to reduce cardiac weight in the normal rabbit [406]. Furthermore, although propranolol did not prevent the development of hypertension, it did minimize cardiac hypertrophy in experimental renal hypertension [121]. Propranolol was not effective in reducing the hypertension and hypertrophy in spontaneously hypertensive rats or in preventing the hypertrophy induced by minoxidil administration [384]. Also, *Tarazi and Sen* [384] observed that methyldopa did not reduce the cardiac hypertrophy induced by experimental renal hypertension if the blood pressure remained elevated.

Tarazi and *Sen* [384] pointed out that despite evidence suggesting that catecholamines initiate cellular hypertrophy via adenyl cyclase stimulation [320] one cannot conclude that catecholamines mediate all types of cardiac hypertrophy. They further cited reports indicating that sufficient stress can induce cardiac hypertrophy even after reduction of cardiac catecholamines [28, 66]. They concluded that catecholamines play an important role in the development of cardiac hypertrophy resulting from stress and in the reduction of hypertrophy when the hemodynamic load is reduced.

It appears that the degree of depletion of cardiac norepinephrine concentration occurring in some types of moderate hypertrophy depends mainly

Table XVI. Ventricular norepinephrine in hypertensive rats

Group	Body weight	Ventricular weight		Ventricular NE	
	g	mg	mg/g	ng	ng/g
1 WKY[a] (n=11)	363	930	2.56	474	519
SE ±	6.5	30	0.07	25	37
2 SHR[b] (n=18)	329	1,108	3.37	564	520
SE ±	4.2	32	0.07	36	37
3 RHR[c] (n=11)	361	1,158	3.26	550	477
SE ±	16	15	0.14	40	36
p 1 vs 2	<0.001	<0.001	<0.001	<0.05	n.s.
1 vs 3	n.s.	<0.001	<0.001	>0.05	n.s.
2 vs 3	>0.05	n.s.	n.s.	n.s.	n.s.

[a] Normotensive controls (Wistar-Kyoto).
[b] Spontaneously hypertensive rats (Okamoto-Aoki strain from Taconic Farms, Inc.).
[c] 2-kidney Goldblatt hypertension (8 weeks duration) produced in Wistar-Kyoto rats.
The differences in catecholamine concentration among the three groups were not statistically significant; there was a tendency, however, for the hypertrophied ventricles of RHR to have a lower concentration than normal (477 vs 519) as expected in other types of experimental hypertrophy, whereas the hypertrophied heart of SHR did not show the expected reduction in catecholamine concentration (520 vs 519).
From *Tarazi and Sen* [384]; reprinted with permission.

on 'dilution' of sympathetic nerves in the increased cardiac mass; however, with a marked degree of hypertrophy both concentration and content may be profoundly reduced due to defects in synthesis, uptake and binding as well as 'dilution' of sympathetic nerves [384].

On the other hand, in the cardiac hypertrophy accompanying experimental renal hypertension [172, 231] and in the hypertrophy occurring in spontaneously hypertensive rats, the concentration of cardiac catecholamines was reported to be normal (table XVI) [384]. *Tarazi and Sen* [384] found ventricular norepinephrine concentration to be inversely correlated with ventricular weight in both normal and spontaneously hypertensive rats (fig. 30) and in rats with renovascular hypertension (induced by unilateral renal artery constriction in the two-kidney rat model). Reversal of renal hypertension and cardiac hypertrophy by unilateral nephrectomy resulted in an increase of cardiac catecholamines (fig. 31). In contrast, the reversal of hypertension and cardiac hypertrophy in the spontaneously hypertensive rat,

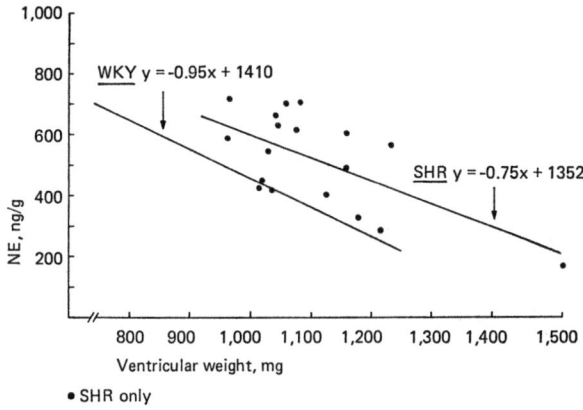

Fig. 30. Ventricular norepinephrine concentration was inversely related to ventricular weight both in normotensive WKY (p < 0.01) and in weight-matched SHR (p < 0.01); the slopes of the regression lines were not significantly different from each other, but the regression line for SHR was significantly displaced upwards compared to normotensive controls. Of particular significance is the observation that the hypertrophied ventricles in SHR did not show the reduction in catecholamine concentration found in experimental pressure hypertrophy (fig. 28, 31). From *Tarazi and Sen* [384]; reprinted with permission.

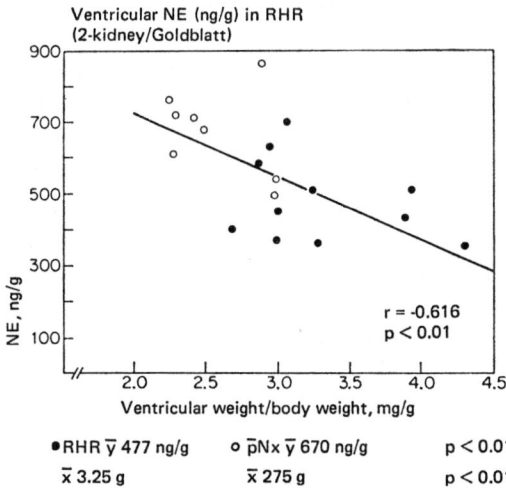

Fig. 31. Correlation of ventricular weight with ventricular NE concentration before and after reversal of hypertension by nephrectomy (pNx) allowed the recognition of subtle changes in cardiac catecholamines in renovascular hypertension (RHR). Ventricular NE concentration was lower in hypertrophied hearts of RHR than following reversal of hypertrophy after cure of hypertension. From *Tarazi and Sen* [384]; reprinted with permission.

Table XVII. Human myocardial tissue

Group	Norepinephrine, µg/g	Epinephrine, µg/g
Right auricle		
Without failure	1.29±0.13	0.11±0.03
With failure	1.06±0.18	0.02±0.00
Left auricle		
Without failure	0.77±0.08	0.06±0.02
With failure	0.43±0.08	0.01±0.00

Table compiled from data of *Borchard* [30]. Values represent mean concentrations of catecholamines (µg/g wet weight). From *Manger* [254]; reprinted with permission.

by administration of antihypertensive drugs, was accompanied by a reduction of cardiac norepinephrine (fig. 29) [384].

As summarized by *Tarazi and Sen* [384], the degree of cardiac hypertrophy and the stimulus inducing it are important in determining cardiac catecholamines and their response to therapy. Of particular interest is the fact that some forms of cardiac hypertrophy and catecholamine abnormalities associated with hypertension are reversible with appropriate therapy.

Borchard [30] reported an interesting study on the adrenergic nerves of normal and hypertrophied hearts of experimental animals and humans which combined biochemical, histochemical, electron microscopic and morphometric analyses. At the time of cardiac operations, specimens were obtained from the auricles of 46 patients with a variety of diagnoses and varying degrees of cardiac hypertrophy. The mean concentrations of catecholamines in these specimens are given in table XVII. A statistically highly significant decrease in norepinephrine content with increasing hypertrophy of muscle fibers was evident. In some cases of moderate hypertrophy, norepinephrine concentrations were not depressed; however, with severe hypertrophy, norepinephrine levels decreased to about 30% of 'control values' (not more than 15% of this loss could be attributed to scarring). With a norepinephrine concentration below 0.6 µg/g wet weight, clinical heart failure was invariably present. *Borchard* suggested that there may possibly be an adaptive augmentation of the adrenergic nerves in the early phase of hypertrophy but that subsequently in the later chronic phase the nerves become less dense and reduced in number. Signs of nerve degeneration also become apparent but the cause for this remains obscure.

In a recent elegant thesis by *Östman-Smith* [300] the conclusion was drawn that: 'The adaptive cardiac hypertrophy produced by chronic exercise is not caused by a direct effect of the increased workload on the cardiac muscle cell, but is instead mediated by release of noradrenaline from cardiac sympathetic nerves. Furthermore, increased activity of cardiac sympathetic nerves may be the final common pathway in all forms of compensatory cardiac hypertrophy.'

A most exciting area of research, currently being pursued by *Kaye* et al. [210], is the production of hypertrophic cardiomyopathy in dogs by the administration of nerve growth factor (a protein necessary for the growth and maintenance of sympathetic nerves). Subcutaneous administration of nerve growth factor to newborn puppies or to pregnant dogs resulted in a marked increase in myocardial norepinephrine and myocardial hypertrophy in the absence of hypertension.

F. The Hyperkinetic 'Beta-Adrenergic' Syndrome and Hypertension

The hyperdynamic beta-adrenergic circulatory state had been defined as a condition in which a hyperkinetic circulation with increased cardiac output (with or without hypertension) results from beta-receptor hyper-responsiveness [129, 131]. Usually there is mild tachycardia at rest; however, these patients periodically have substantial increases in their heart rate (even up to 170 bpm [129]) with palpitations and, not infrequently, attacks of apprehension, sometimes with severe anxiety and dyspnea. These attacks may be precipitated by emotional excitement, standing, physical exertion, or administration of isoproterenol [131]. Chest and abdominal discomfort and orthostatic hypotension have been noted [129]. Sustained systolic and/or diastolic hypertension of mild to moderate severity may be present. Frequently the blood pressure may be labile and become elevated during episodes of tachycardia.

The findings of an elevated cardiac output and increased heart rate [24, 104, 130, 201, 249, 340, 343] in a significant number of subjects with 'hyperkinetic' borderline hypertension has further prompted the question of whether this augmented output leads to established hypertension by triggering an increased peripheral resistance via the mechanism of autoregulation [161, 202].

To test the hypothesis that an increased cardiac performance can lead to hypertension, *Liard* et al. [240] studied the hemodynamic and humoral characteristics of acute hypertension induced by prolonged (7 days) stellate ganglion stimulation, in conscious dogs. An abrupt rise in mean arterial pressure occurred which was entirely due to increased cardiac output. However, following only 1 day of stimulation, the cardiac output had returned to control values and hypertension was sustained entirely by increased peripheral resistance. These hemodynamic effects were apparently mediated by increased activity of the sympathetic nervous system. When stimulation was discontinued, arterial pressure returned to control levels. Unfortunately, this model differed from what was anticipated; the increased cardiac output was very brief and the hypertension was not accompanied by a hyperkinetic circulation with an increased heart rate. It was concluded that in this particular model, hypertension may not have resulted from autoregulation but may have been caused by stimulation of a pressor reflex.

It has been suggested that in about 70% of borderline hypertensives with a normal cardiac output, increased peripheral resistance could be attributed to nonneurogenic structural changes of the vessels which impede arterial blood flow; however, in approximately 30% of these patients, enhanced alpha-adrenergic vasomotor tone was evident [70, 110, 204]. An increased venous tone in borderline hypertension has also been reported but the mechanism for this latter abnormality is uncertain, since attempts to demonstrate an augmented alpha-adrenergic venomotor activity in this condition have failed [204].

In borderline hypertension [110, 281] and 'hyperkinetic' borderline hypertension [98, 281] a substantial number of patients (30% in one series [110]) have elevated plasma renin levels at rest.

Some patients with borderline hypertension have a more pronounced increase in plasma renin than normal subjects with head-up tilting or standing [111, 225, 280, 415]; furthermore, in some patients, standing or tilting caused excessive excretion of urinary norepinephrine which was positively correlated with the increase in plasma renin [111, 225]. Because of the importance of the renal sympathetic nerves and catecholamines on renin secretion [397], the question has been raised as to whether a heightened sympathetic nerve activity is responsible for elevated plasma renin concentrations in borderline hypertension and in 'hyperkinetic' borderline hypertension [204].

Julius et al. [204] concluded that in patients with borderline hypertension who have an increased adrenergic vasomotor tone and an elevated level of plasma renin, the hypertension results from increased adrenergic activity. Their conclusion was based on the demonstration that beta-adrenergic blockade with propranolol caused a substantial fall in plasma renin without an appreciable effect on the blood pressure, whereas alpha-adrenergic blockade with phentolamine caused an abrupt decrease in blood pressure and no further reduction in plasma renin. These patients were also found to have an increased cardiac drive since propranolol administration caused a greater reduction of heart rate in borderline hypertensives than normal subjects [203].

It was postulated that in the majority of borderline hypertensives, there is abnormal integration of autonomic nervous control, probably originating in the medulla oblongata [204]. The enhanced adrenergic activity is generalized and involves both alpha and beta receptors. In addition, they presented evidence of a diminished parasympathetic inhibition of the heart in patients with 'hyperkinetic' borderline hypertension. As pointed out by these investigators this reciprocal relationship of the autonomic nervous system (i.e. an increased sympathetic drive and a diminished parasympathetic inhibition) is characteristic of the functional organization in the higher integrative areas of autonomic control [204].

G. Cardiac Pathophysiology Associated with Excessive Adrenergic Activity or Excessive Circulating Catecholamines

The effects and significance of excessive adrenergic activity in arrhythmias, myocardial ischemia, and heart failure have already been discussed. A number of clinical conditions are associated with augmented adrenergic activity, and it is conceivable that, if the activity is profoundly increased and/or prolonged, the electrophysiology and function of the heart may be significantly influenced. Furthermore, cardiopathology may appear.

A fascinating recent experimental development is the demonstration by *Witzke* and *Kaye* [425] that nerve growth factor (a glycoprotein) enhances sympathetic nerve growth, increases cardiac adrenergic innervation, and elevates myocardial norepinephrine. When nerve growth factor was administered to newborn puppies, changes in structure and function occurred which were similar to those seen in human idiopathic hypertrophic subaortic stenosis. The hypothesis that nerve growth factor might be involved in the pathogenesis of this disease is a most intriguing idea requiring confirmation.

By far the most informative condition for studying the effects of excessive circulating catecholamines on the heart is pheochromocytoma. In this disease, concentrations of circulating catecholamines are often markedly

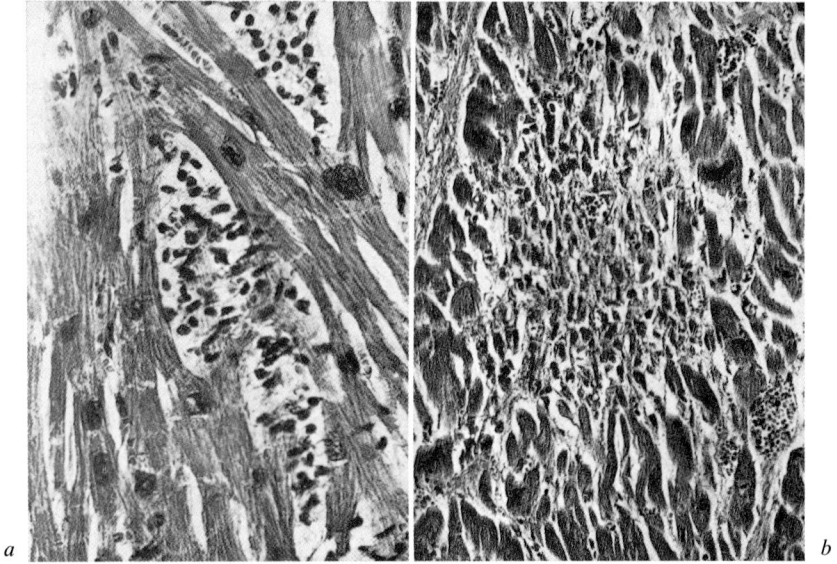

Fig.32. a Left ventricular myocardium of a 45-year-old man, who died with an unsuspected pheochromocytoma 5 days after removal of the left kidney, showing active myocarditis, with focal degeneration of myocardial fibers and inflammation. Hematoxylin and eosin. × 375. Delicate collagenous fibers were present in these foci. (This patient's ECG is shown in fig.33.) *b* Left ventricular myocardium of a 38-year-old man, who died of congestive heart failure that was thought to be due to a myocarditis. Hemotoxylin and eosin. × 200. An unsuspected pheochromocytoma was present. The myocardium showed active myocarditis consisting of foci of degenerated myocardium fibers and inflammation. From *Van Vliet* et al. [400]; reprinted with permission.

elevated. In addition to the large number of pathologic and metabolic complications caused by circulating catecholamines, chronic exposure of the heart to elevated plasma concentrations of catecholamines can cause a cardiomyopathy and pronounced electrophysiological alterations [255].

1. Cardiomyopathy in Pheochromocytoma

Cardiomyopathy has been noted at autopsy in some patients who died from complications of pheochromocytoma [218, 332, 400]. *Van Vliet* et al. [400] reported that of 26 patients at the Mayo Clinic who died with pheochromocytoma, 15 (58%) had active myocarditis which these authors believed resulted from excessive catecholamines. Examples of active myocarditis in 2 of their patients is evident in figure 32. Focal areas of degeneration and necrosis of myocardial fibers, with foci of inflammatory cells (predominantly

histiocytes, but some plasma cells and occasional polymorphonuclear leuko-
cytes) were present and most numerous in the left ventricle. In addition,
increased fibrosis in these foci and diffuse edema of the myocardium were
observed. In a few cases of hearts with active myocarditis, there was thickening of small and medium-sized arteries due to edema of the intima and media,
and some fibrous replacement of the media. A moderately severe degree of
coronary sclerosis was found in 14 of the 26 patients.

The pathologic changes of active myocarditis were found in patients
with pheochromocytoma who had sustained or paroxysmal hypertension.
The lesions were similar to those seen in the myocardium of some patients
who had received therapeutic infusions of norepinephrine, and to lesions
found in various laboratory animals after injections of catecholamines. It is
interesting that oxidation products of catecholamines such as adrenochrome
may be implicated in the pathogenesis of experimental catecholamine cardiomyopathy [438]. No active catecholamine myocarditis was evident in
4 patients with pheochromocytomas which were apparently nonfunctioning
[400].

11 of the 15 patients with active myocarditis had left ventricular failure
with pulmonary congestion, which was frequently accompanied by a period
of hypotension (refractory to treatment) prior to death. In 1 patient a
diagnosis of active myocarditis had been made on the basis of cardiomegaly,
dyspnea, tachycardia, and palpitations. Signs and symptoms had been
present for 13 years before a pheochromocytoma was discovered at autopsy
[400]. *Engelman and Sjoerdsma* [106] reported a patient whose cardiomyopathy, presumably related to excess catecholamine, subsided after
resection of the pheochromocytoma. Significant left ventricular failure may
occur in the absence of longstanding or severe hypertension [435].

2. Electrocardiographic Changes in Pheochromocytoma

(a) *Clinical.* Electrocardiographic changes, particularly sinus tachycardia, may be frequently observed in patients with pheochromocytoma.
Atrial or ventricular premature contractions or tachycardia are not infre-
quent during hypertensive crises. Supraventricular tachycardia up to 200/min
in 2 patients and 160/min in a third patient has been recorded [342]. Ectopic
beats may result from stimulation of the myocardium by increased circulating catecholamines [96]. Furthermore, myocardial 'irritability' is increased if
its concentration of catecholamines is excessive. In their review of alterations
in the electrocardiogram of patients with pheochromocytoma, *Sayer* et al.
[346] have categorized these as abnormalities of rhythm or abnormalities

suggesting myocardial ischemia, damage, or strain. Disorders of rhythm consisted of the following: (1) wandering pacemaker; (2) sinoauricular dissociation; (3) auricular tachycardia; (4) auricular premature contractions; (5) auricular flutter; (6) auricular fibrillation; (7) nodal tachycardia; (8) ventricular premature contractions, and (9) ventricular tachycardia.

They found that arrhythmias occurred most frequently during paroxysms of hypertension and sometimes persisted after the blood pressure had returned to normal levels. All reported arrhythmias disappeared after the tumors were removed [346]. Following administration of benzodioxane or phenoxybenzamine to 1 of their patients with pheochromocytoma and sustained hypertension, *Sayer* et al. [346] observed a reversion toward normal of T waves, prolongation of the Q-T interval, and a clockwise rotation of the heart, with a shift to a more vertical position. ECG abnormalities suggesting myocardial damage included the following: (1) Left axis deviation; (2) Right axis deviation (occasionally with patterns similar to those seen in acute cor pulmonale and related to a marked increase in pulmonary artery pressure and a disproportionate increase in right ventricular work, which can be produced by catecholamine infusions [424]; (3) Abnormally high or peaked P waves; (4) Low or inverted T waves (often diffusely distributed); (5) S-T segment deviations and prolongation of Q-T interval.

Sayer et al. [346] reported that these changes occurred transiently during or between paroxysms of hypertension, and continuously in some patients with sustained hypertension. They stated that disappearance of these abnormalities sometimes occurred spontaneously, during hypertension paroxysms or during a low-sodium regimen. Partial or complete reversal of these abnormalities to normal was noted in patients after tumor removal. It must be kept in mind, however, that evidence of irreversible myocardial damage [e.g. that associated with catecholamine cardiomyopathy as described by *Kline* [218] and *Van Vliet* et al. [400] or with coronary atherosclerosis and/or myocardial infarction] may persist after a hypertensive crisis caused by pheochromocytoma or after tumor removal.

Sayer et al. [346] emphasized that in the absence of any demonstrable etiologic factors the ECG abnormalities described above should alert the clinician to consider the possibility of pheochromocytoma. They appropriately stated that: 'The pathogenesis of the abnormal electrocardiogram produced by a pheochromocytoma is a complex interplay of the relative amounts of epinephrine and norepinephrine secreted by the tumor, the duration of the secretion, whether intermittent or sustained, and the net effects of these pressor amines upon the cardiac rate, rhythm, output, oxygen demand and supply, as well as the coronary circulation, pulmonary and peripheral arterial resistance, and perhaps the body electrolyte distribution.' It is well to remember, as indicated by *Futterweit* et al. [133] and noted by others, that a singularly striking feature of the ECG of some patients with pheochromocytoma is a diffuse distribution of T and S-T changes [46, 189, 346, 354]. These changes are in contrast to those observed in patients with other forms of hypertension exhibiting a 'strain pattern'.

More recently, *Saint-Pierre* et al. [342] have reviewed the electrocardiographic findings in patients with pheochromocytoma reported in the literature and have also presented the findings in 21 of their own patients. They noted that ventricular extrasystoles were particularly common and were recorded in about 50% of the reports in the literature. Auricular extrasystoles occurred less frequently. Of the patients reported by *Saint-Pierre* et al. [341], the following ECG abnormalities were noted: (1) left ventricular hypertrophy patterns; (2) arrhythmias (occurring sometimes without hypertension and liable to be induced by effort); (3) coronary insufficiency patterns; (4) variable and transient repolarization disturbances (probably related to functional or autonomic disturbances caused by high concentrations of blood catecholamines); (5) disturbances of repolarization accompanying clinical, roentgenologic, and hemodynamic manifestations of myocardial involvement (catecholamine cardiomyopathy).

None of the changes in the ECG described above can be considered specific; but, when they occur in association with the onset of palpitations, hypertension, and other symptoms and signs of increased circulating catecholamines, they take on greater diagnostic significance. The extent to which an associated coronary atherosclerosis might have contributed to some of the ECG abnormalities encountered is difficult to assess, especially since the effects of pheochromocytoma may possibly facilitate coronary atherogenesis [309].

A particularly fascinating electrocardiographic pattern, consistent with that observed in acute anterior myocardial infarction (note S-T elevations in precordial leads and Q wave in aVL), was observed in one of the patients with pheochromocytoma studied at the Mayo Clinic (fig. 33). These ECG changes apparently were not the result of myocardial infarction but were probably related to the effects of excessive circulating catecholamines.

Electrocardiographic changes consistent with acute anteroseptal myocardial infarction have recently been reported in another patient with an adrenal pheochromocytoma studied at the Mayo Clinic (fig. 34). This 59-year-old woman never experienced chest pains, and because of rapid reversal

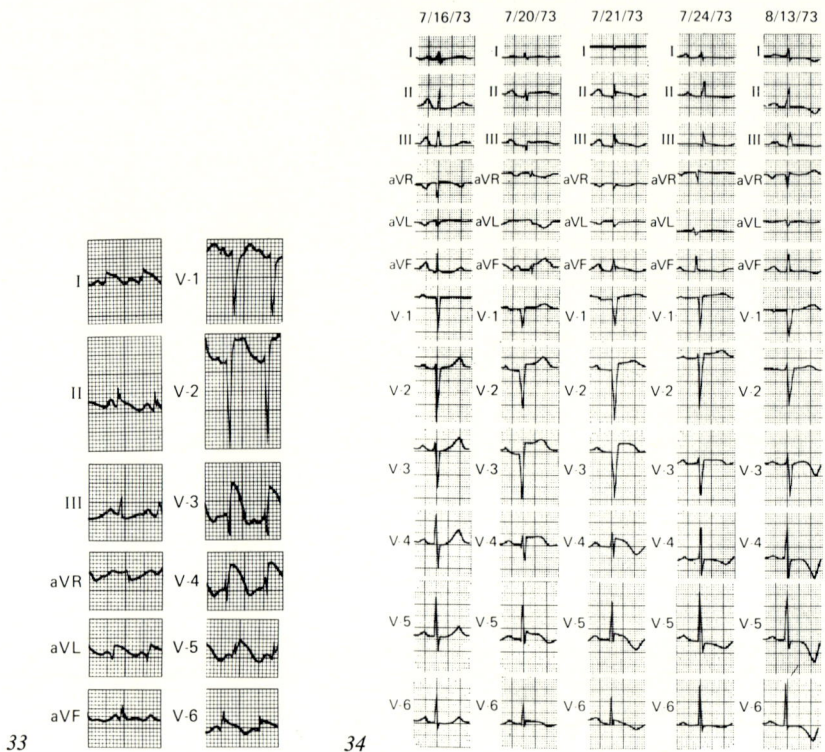

Fig. 33. This ECG was recorded about 24 h before death in a 45-year-old man with a right adrenal pheochromocytoma who was in circulatory shock. He had not been recognized as having a pheochromocytoma and, following a left nephrectomy for transitional cell carcinoma of the kidney pelvis, was found to be in circulatory collapse. Postmortem examination revealed no evidence of acute myocardial infarction, no mural thrombosis, and no significant coronary artherosclerosis; an incidental finding was an atrial septal defect. However, scattered throughout the myocardium were focal areas of necrosis associated with myocardial fiber degenerative changes and inflammatory cell infiltrations. It was felt by Dr. *A. L. Brown* (pathologist at the Mayo Clinic) that these pathologic legions could best be explained on the basis of excessive circulating catecholamines from his pheochromocytoma. (This patient's myocardial pathology is shown in fig. 32a.) Courtesy of Dr. *H. B. Burchell*, Senior Consultant in Cardiology, University of Minnesota. From *Manger and Gifford* [255]; reprinted with permission.

Fig. 34. Serial electrocardiograms. The S–T segment elevation and loss of R-wave voltage over anterior precordial leads on July 20, 1973 suggested acute anteroseptal infarction. There is S–T segment elevation in leads II, III, and aVF. Note also the rapid resolution of these changes: by July 24, 1973, R-wave voltage was returning to normal and by August 13, 1973, it was normal and only widespread T-wave inversion was evident.

of the changes, they were attributed to a toxic myocarditis due to catechol-amines, rather than to a transmural infarction [321]. Others have reported a similar case [308].

Cheng and *Bashour* [56] have recently reported the case of a 51-year-old black woman with pheochromocytoma who presented with striking cardio-graphic changes mimicking ischemic heart disease at one time and acute pulmonary embolism at other times. Paroxysmal episodes were accompanied by abnormalities of repolarization of the Q-T interval and deep and wide symmetrically inverted T waves without changes in the QRS complex. Pro-minent P waves in leads II, III, and aVF were frequently observed with no significant change in the QRS axis. The Q-T interval remained prolonged even when T waves were upright. Normal ECGs were recorded during several attacks of paroxysmal hypertension. These investigators pointed out that symmetric T wave inversion with or without a prolonged Q-T interval is not specific for coronary artery disease; similar ECG changes have been seen in acute pancreatitis, gallbladder disease, acute cerebrovascular accidents, during administration of quinidine, atypical angina with normal coronary arteriograms, and diencephalic discharges. Although the mechanism whereby these changes induce repolarization abnormalities is not well understood, a diencephalic discharge of catecholamines has been implicated in some of these conditions. Finally, *Cheng* and *Bashour* [56] cite evidence for the toxic effects of excessive levels of catecholamines on the myocardium. They also noted that reversible T wave inversions have been induced in normal man by intravenous infusion of norepinephrine.

It is interesting that electrocardiographic changes similar to those which occurred in the patient of *Cheng* and *Bashour* [56] were observed in a patient who inadvertently received an overdose of norepinephrine and developed circulatory shock [*Burchell*, personal communication].

The ECG was that of a 59-year-old white woman who developed a paroxysm of hyper-tension (250/105) and became apathetic, disoriented and transiently blind. A left hemi-sensory deficit, dysesthesia of the left leg, and a left Babinski were noted and shortly thereafter she became unresponsive, diaphoretic, and showed signs of peripheral vaso-constriction. The ECG became abnormal at this time and thereafter sequential changes were observed as shown in the figure. She regained her sight in a few hours but peripheral vasoconstriction persisted and the sensory deficits improved only gradually. She was treated with alpha- and beta-adrenergic blocking agents, and subsequently a pheochromo-cytoma was successfully removed from the region of the left adrenal gland. From *Radtke* et al. [321]; reprinted with permission.

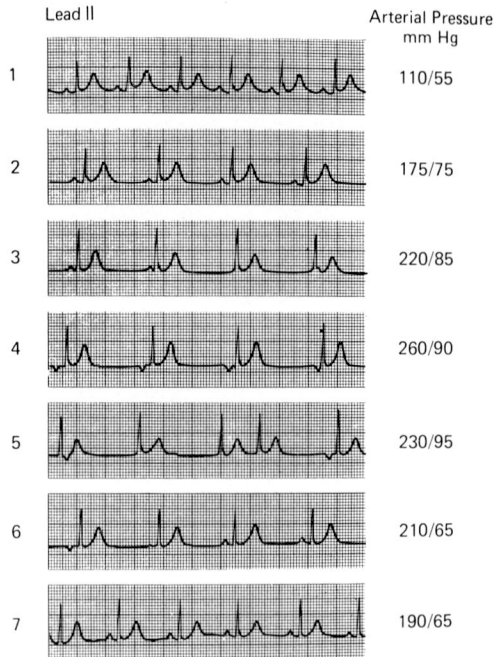

Fig.35. Sequential changes in cardiac rhythm during an episode of hypertension in case 1, demonstrating slowing of the sinus pacemaker and nodal escape rhythm with A–V dissociation as the blood pressure rises, and restoration of sinus rhythm, culminating in sinus tachycardia, as the blood pressure falls toward normal (paper speed 25 mm/s). Note the initial drop in blood pressure (shown in table XVIII) at the onset of this attack when the epinephrine became elevated. On one occasion the blood pressure rose from 120/80 to 250/100 mm Hg during the development of A–V dissociation and then fell to 80/60 mm Hg. This is particularly interesting since hypotensive periods have occasionally been reported in patients harboring a pheochromocytoma which secretes predominantly epinephrine. This is patient No. 30 described by *Manger and Gifford* [255]. Numbers on the left-hand margin indicate time in minutes. From *Forde* et al. [124]; reprinted with permission.

Occasionally bradycardia instead of tachycardia occurs in patients with pheochromocytoma [124, 164, 171, 364]. Vagal response to elevated blood pressure during hypertensive crises in some of these patients evidently causes sinus slowing instead of the usual sinus tachycardia, premature beats, and ectopic tachycardias which result from β-adrenergic stimulation of the myo-cardium [124]. Rarely, reflex bradycardia with depression of the sinus node and escape of lower pacemakers have been recorded in patients with pheo-

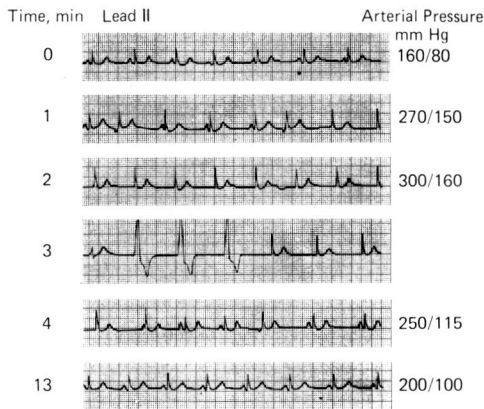

Fig. 36. Sequential changes in cardiac rhythm during an episode of hypertension precipitated by massage of right upper abdominal quadrant in case 2. Note depression of sinus pacemaker and appearance of lower foci as blood pressure rises. When blood pressure declines, sinus mechanism is restored (paper speed 25 mm/s). This is patient No. 32 described by *Manger and Gifford* [255]. From *Forde* et al. [124]; reprinted with permission.

chromocytoma during a severe hypertensive crisis [44, 112, 124]. Nodal escape and A-V dissociation have been observed during hypertensive crises in association with both bradycardia [44, 112, 124] and tachycardia [133].

Forde et al. [124] have reported 2 patients with intermittent chest pain and arrhythmia who were initially suspected of having acute myocardial infarction. Constant ECG and blood pressure monitoring revealed that episodically severe hypertension (associated with anterior chest pressure or pain, palpitations, and other clinical manifestations) occurred coincidentally with reflex bradycardia and nodal escape rhythm and A-V dissociation (fig. 35, 36). With subsidence of the hypertensive crises, the ECG reverted to normal sinus rhythm. The rise in blood pressure always preceded the onset of the arrhythmia. In case 1 (fig. 35) it was found that only the plasma epinephrine became markedly elevated during a spontaneous hypertensive crisis, whereas in case 2 (fig. 36) both epinephrine and norepinephrine plasma concentrations reached very high levels when hypertension was induced by massage of the right upper abdomen (table XVIII). It is noteworthy that sinus bradycardia and occasionally complete A-V dissociation have been produced by intravenous infusion of norepinephrine into human subjects [21].

As emphasized by *Forde* et al. [124], the diagnosis of pheochromocytoma should be considered in any patient with periodic bradycardia and escape of

Table XVIII. Plasma levels of epinephrine and norepinephrine in 2 cases of pheochromo-cytoma with reflex bradycardia and nodal escape rhythm[a]

	Blood pressure mm Hg	Plasma epinephrine[b] µg/l plasma	Plasma norepinephrine[b] µg/l plasma
Normal concentrations (upper limit)		< 1.5	< 6.6
Case 1			
Control	154/ 80	1.3[c]	3.8[c]
During spontaneous hypertensive attack			
Onset	80/ 50	2.6	3.5
Rising	205/ 68	26.7	3.6
Peak	240/130		
Falling	198/ 73	2.9	3.4
Case 2			
Control	158/ 80	3.1[c]	8.2[c]
During hypertensive attack induced by pressure in right upper quadrant			
Peak	300/160	28.5[c]	27.3[c]

[a] Modified from *Forde* et al. [124].
[b] Performed by Dr. *W. M. Manger*, New York University Medical Center.
[c] Average of two rapid sequence samples.
From *Manger and Gifford* [255]; reprinted with permission.

lower pacemakers. Careful documentation of the blood pressure before and during such arrhythmias is mandatory, and if it is disclosed that these electrocardiographic abnormalities are induced by hypertensive crises, a pheochromocytoma must be strongly suspected. To date, reflex bradycardia with nodal escape rhythm in patients with pheochromocytoma has almost invariably been described in those who have paroxysmal and not persistent hypertension. One 44-year-old woman with slight hypertension (160/108 mm Hg recumbent) had isoelectric T waves in leads I and aVL, diphasic T waves in leads V3–4, and inverted T waves in leads II, III, and V5–6. Slight S-T depression in leads V5–6 and a prominent P wave in lead II were also present [133]. She developed a blood pressure rise of 80/46 mm Hg above the basal level following a histamine (0.025 mg i.v.) provocative test, and simul-taneously the ECG revealed further peaking of the P waves, A-V dissocia-tion, and supraventricular and ventricular extrasystoles. The ECG returned to the prestimulation state 20 min after histamine administration. 3 days

a Onset of attack (BP = > 300/115, normal BP = 130/65).
Nodal escape rhythm—heart rate = 58/min

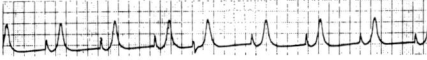

b Symptoms subsiding. Nodal escape with some sinus
rhythm—heart rate = 51-63/min

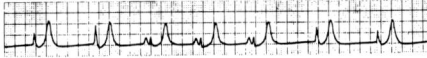

c 3-5 minutes after onset, symptoms gone (BP = 135/60).
Regular sinus rhythm—heart rate = 82/min

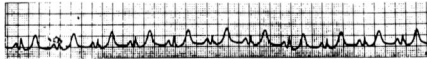

Fig. 37. Continuous recording from lead II (coronary care unit monitoring electrodes) in a 58-year-old white male with pheochromocytoma during paroxysmal attack of hypertension. This is the same patient whose ECG appears in figure 36. From *Manger and Yormak*, p. 190, in *Manger and Gifford* [255]; reprinted with permission.

following pheochromocytoma extirpation no ECG abnormalities were evident.

An electrocardiographic finding which has been noted [44, 128, 310] in patients with pheochromocytoma, but not previously emphasized, was observed in case 2 mentioned above [*Manger* and *Yormak*, unpublished data]. Figure 37 reveals a continuous recording from lead II in this patient during a paroxysmal attack of hypertension. During the periods of bradycardia and nodal escape, the T waves became exceptionally high and decreased toward normal when the blood pressure returned to normal and the bradycardia subsided. Also, there was a slight and transient depression of the S-T segment. The finding of a huge T wave is in no way specific for pheochromocytoma. Enlarged T waves may be seen in myocardial infarction or hyperkalemia; however, one would usually expect other typical ECG changes to occur concomitantly with these latter conditions.

The ECG of a patient with pheochromocytoma reported by *French and Campagna* [128] revealed a prolonged Q-T interval, prominent U waves, depressed S-T segments, and high, peaked T waves – abnormalities that were clearly reversible.

Table XIX. ECG abnormalities sometimes seen in patients with pheochromocytoma

---

Atrial, nodal, or ventricular tachycardia
Atrial or ventricular premature contractions (with or without bigeminy)
Bradycardia with or without complete A–V dissociation with or without T wave peaking
Wandering pacemaker
Atrial flutter or fibrillation
Left or right axis deviation
Abnormally high or peaked P waves
Low or inverted T waves
S–T segment deviations
Prolonged Q–T interval
(Reversibility of ECG abnormalities + absence of etiologic explanation suggests pheochromocytoma)

From *Manger and Gifford* [255]; reprinted with permission.

A summary of the electrocardiographic abnormalities which may be seen in patients with pheochromocytomas which are actively secreting catecholamines is given in table XIX.

(b) *Experimental. Tenzer* [385] has stated that epinephrine injected into the normal man increases cardiac frequency and the amplitude of the P waves, shortens the P-R interval, increases myocardial excitability, and provokes alterations in rhythm. In addition, he mentions that therapeutic doses of epinephrine can cause elevation of T waves, whereas larger doses will cause T wave inversion. When moderate doses of norepinephrine were injected, bradycardia and depression of the P wave occurred; in addition, nodal rhythm sometimes appeared, and frequently the T waves became elevated.

It is noteworthy that infusions of epinephrine or norepinephrine into the coronary artery of dogs produced high, upright, peaked T waves with depressed S-T segments [19]. Large upright T waves, associated with prominent U waves, prolonged Q-T interval and T-U fusion have not infrequently been observed as an electrocardiographic manifestation of intracranial disease [43].

Experimental animal studies on the effect of hypothalamic or stellate ganglion stimulation, and stellate ganglionectomy, suggest that sympathetic pathways are involved in some of these ECG alterations [277, 436]. Increased concentrations of circulating catecholamines may conceivably cause disturbances in the myocardium in a manner similar to those induced by activation of the sympathetic innervation of the heart. Intravenous administration of epinephrine or norepinephrine to healthy subjects leads to reversible ECG alterations. *Sjöstrand* [360] found that epinephrine invariably caused the appearance of a positive after-potential following the T wave which was coupled with an associated depression of the T wave and S-T segment. After vagal block the ECG changes caused by epinephrine were augmented. On the other hand, administration of norepinephrine caused the reverse of the epinephrine effect; i.e. norepinephrine in the same dose produced a slight rise in the T wave, while the positive after-potential was either decreased or unchanged and the heart rate was slowed. ECG changes similar to those evoked by epinephrine were encountered in 7 of 8 patients with pheochromo-cytoma by *Cannon and Sjöstrand* [49]. They emphasized that marked and variable reversible ECG changes occurring in a very short time should suggest the possibility of pheochromocytoma.

Watkins [411] had commented that ECG alterations could be related to shifts in cellular potassium accompanying hyperglycemia caused by increased circulating catecholamines. Consistent with this concept, *French and Campagna* [128] suggested that altered potassium metabolism, with subsequent development of hypokalemia, could account for prolonged Q-T intervals, S-T depression and prominent U waves in addition to the positive after-potential (where the U or T wave descends and merges with the P wave before reaching the baseline) noted in their patient and described by *Cannon and Sjöstrand* [49]. They further pointed out that although the upright, peaked T wave in their patient was not consistent with hypokalemia, cate-cholamines can secondarily cause a marked vagal effect, which could account for the T wave peaking [48]. *Burchell* [personal communication], who kindly reviewed this manuscript, suggested the possibility that a disparate flow of catecholamines into the myocardium might occur and result in local differ-ences in sympathetic 'drive' which could produce large T waves; he felt, however, that a transient ischemia of the subendocardial zone was a more likely explanation of the huge T waves.

It should be pointed out that concentration of serum potassium during paroxysms of hypertension in patients with pheochromocytoma have not been adequately studied. During a constant intravenous infusion of epineph-rine (10–18 µg/min) in 12 normal young men, the mean serum potassium decreased very significantly by 17.3% without any remarkable change in serum sodium or chloride [193]. Concomitantly, there was a decrease in urinary sodium, chloride, and potassium. The alteration in serum potassium was thought to reflect an intracellular shift. *Keys* [213] noted a similar decrease in potassium in normal men following intravenous injection of 0.005–0.3 mg of epinephrine chloride but a return to levels slightly above preinjection values within 40–60 min. More recently, *Massara* et al. [270] demonstrated that intravenous infusion of epinephrine (0.01 mg/min) into healthy volunteers caused approximately a 20% decrease in serum potassium which could be prevented by propranolol. The latter finding suggested that epinephrine-induced hypokalemia was mediated by beta-adrenergic receptors. These findings were in contrast to the immediate marked rise in plasma potassium caused by intravenous injections of epinephrine in dogs, cats, and rabbits [213]. A number of other investigators have found that administration of epinephrine causes a decrease in serum potassium [38, 53, 97, 329].

Graded intravenous infusions of norepinephrine in 6 normal young men caused a sharp fall in potassium clearance by the kidney without remarkably affecting clearance of sodium chloride [317]; also, there was a slight but statistically significant rise in serum potassium concentration (which averaged 5.2%) without any remarkable change in serum sodium or chloride. These results suggested that norepinephrine might cause a migration of potassium from intracellular to extracellular fluid. *MacKeith* [251], however, reported 1 patient with pheochromocytoma who had very elevated serum potassium between hypertensive attacks; he pointed out that intravenous epinephrine may produce an 86% rise in blood potassium. Only rarely has the serum potassium been reported elevated in patients with pheochromocytomas [273]. Serum electrolytes are nearly always within normal limits in those patients with pheochromocytoma who have sustained hypertension; they are also normal between the hypertensive crises in those patients who have paroxysmal hypertension with pheochromocytoma.

In the final analysis, we must recognize how totally imprecise are the contributions made by the electrocardiogram to the diagnosis of pheochromocytoma. After kindly reviewing this section on electrocardiography, Dr. *Raymond Pruitt* emphasized the desirability of pointing out the near total lack of specificity of the electrocardiographic changes occurring in patients with pheochromocytomas. He remarked that the origin of ECG changes may be more readily identified from clinical phenomena which suggest the presence of pheochromocytoma than from the ECG changes themselves [*Pruitt*, personal communication].

VI. Concluding Remarks

In the above account an attempt has been made to demonstrate clearly that the adrenergic nervous system and circulating catecholamines can play an important role in normal physiology and pathophysiology of the heart. It should also be mentioned that a vast array of experimental physiologic and pharmacologic manipulations, including administration of certain drugs – e.g. drugs which influence adrenergic neural transmission and the synthesis and inactivation of catecholamines – and alterations in sodium [95] and hormonal balance [83] can change the concentrations and turnover rate of catecholamines in the heart and elsewhere [126, 248, 290, 296, 328, 330]; however, space does not permit their consideration in this review.

A recent comprehensive account of current concepts of cardiac function (including neural control of the heart and biochemical mechanisms of adrenergic and cholinergic regulation of myocardial contractility) has been presented elsewhere [27] and may be of additional interest to the reader.

References

1 Abboud, F.M.: Control of the various components of the peripheral vasculature. Fed. Proc. *31:* 1226–1239 (1972).

2 Abildskov, J.A.; Vincent, G.M.: The autonomic nervous system in relation to electrocardiographic waveform and cardiac rhythm; in Randall, Neural regulation of the heart, pp. 409–424 (Oxford University Press, New York 1977).

3 Ahlmark, G.; Saetre, H.; Korsgren, M.: Reduction of sudden deaths after myocardial infarction. Lancet *ii:* 1563 (1974).

4 Alderman, E.I.; Harrison, D.C.: Myocardial hypertrophy resulting from low dosage isoproterenol administration in rats. Proc. Soc. exp. Biol. Med. *136:* 268–270 (1971).

5 Allen, J.D.; Pantrigde, J.F.; Shanks, R.G.: The effects of practolol on the dysrhythmias complicating acute ischemic heart disease. Am. J. Med. *58:* 199–208 (1975).

6 Amorium, D.S.; Mello de Oliveira, J.A.; Manço, J.C.; Gallo, L. Jr.; Meira de Oliveira, J.S.: Chagas heart disease: first demonstrable correlation between neuronal degeneration and autonomic impairment. Acta cardiol. *28:* 431–440 (1973).

7 Anderson, J.; Del Castillo, J.: Cardiac innervation and synaptic transmission in the heart; in Demello, Electrical phenomena in the heart, pp. 236–257 (Academic Press, New York 1972).

8 Armour, J.A.; Hageman, G.R.; Randall, W.C.: Arrhythmias induced by local cardiac nerve stimulation. Am. J. Physiol. *223:* 1068–1075 (1972).

9 Armour, J.A.; Wurster, R.D.; Randall, W.C.: Cardiac reflexes; in Randall, Neural regulation of the heart, pp. 159–186 (Oxford University Press, New York 1977).

10 Arnsdorf, M.F.; Hsieh, Y.-Y.: Pharmacology of cardiovascular drugs; in Hurst, Logue, Schlant, Wenger, The heart; 4th ed., pp. 1943–1963 (McGraw-Hill, New York 1978).

11 Avakian, O.V.; Gillespie, J.S.: The relationship between the development of fluorescence on the response of arterial smooth muscle perfused with noradrenaline. J. Physiol., Lond. *191:* 71p–72p:(1967).

12 Avakian, O.V.; Gillespie, J.S.: Uptake of noradrenaline by adrenergic nerves, smooth muscle and connective tissue in isolated perfused arteries and its correlation with the vasoconstrictor response. Br. J. Pharmacol. *32:* 168–184 (1968).

13 Axelrod, J.: O-Methylation of epinephrine and other catechols in vitro and in vivo. Science *126:* 400–401 (1957).

14 Axelrod, J.: Purification and properties of phenylethanolamine-N-methyl transferase. J. biol. Chem. *237:* 1657–1660 (1962).

15 Axelrod, J.: The formation, metabolism, uptake and release of noradrenaline and adrenaline; in Variey, Gowencock, Symp. on the clinical chemistry of monoamines, pp. 5–18 (Elsevier, Amsterdam 1963).

16 Axelrod, J.: The fate of noradrenaline in the sympathetic neurone; in The Harvey Lectures, series 67, pp. 175–197 (Academic Press, New York 1972).

17 Axelrod, J.; Inscoe, J.K.; Senoh, S.; Witkop, B.: O-Methylation, the principle pathway for the metabolism of epinephrine and norepinephrine in the rat. Biochim. biophys. Acta 27: 210–211 (1958).

18 Axelrod, J.; Kopin, I.J.; Mann, J.D.: 3-Methoxy-4-hydroxyphenylglycol sulfate: a new metabolite of epinephrine and norepinephrine. Biochim. biophys. Acta 36: 576–577 (1959).

19 Barger, A.C.; Herd, J.A.; Liebowitz, M.R.: Chronic catheterization of coronary artery: induction of ECG pattern of myocardial ischemia by intracoronary epinephrine. Proc. Soc. exp. Biol. Med. 107: 474–477 (1961).

20 Barner, B.D.; Jellenik, M.; Kaiser, G.C.: Effects of isoproterenol infusion on myocardial structure and composition. Am. Heart J. 79: 237–241 (1970).

21 Barnett, A.J.; Blacket, R.B.; DePoorter, A.E.; Sanderson, P.H.; Wilson, S.M.: The action of noradrenaline in man and its relation to pheochromocytoma and hypertension. Clin. Sci. 9: 151–179 (1950).

22 Baron, G.D.; Speden, R.N.; Bohr, D.F.: Beta-adrenergic receptors in coronary and skeletal arteries. Am. J. Physiol. 223: 878–881 (1972).

23 Beavo, J.A.; Hardman, J.G.; Sutherland, E.W.: Stimulation of adenosine 3′, 5′-monophosphate hydrolysis by guanosine 3′, 5′-monophosphate. J. biol. Chem. 246: 3841–3846 (1971).

24 Bello, C.T.; Sevy, R.W.; Harakal, C.: Varying hemodynamic patterns in essential hypertension. Am. J. med. Sci. 250: 24–35 (1965).

25 Berne, R.M.: Effect of epinephrine and norepinephrine on coronary circulation. Circulation Res. 6: 644–655 (1958).

26 Berne, R.M.; DeGeest, H.; Levy, M.N.: Influence of the cardiac nerves on coronary resistance. Am. J. Physiol. 208: 763–769 (1965).

27 Berne, R.M.; Sperelakis, N.; Geiger, S.R. (eds): The heart. Handbook of Physiology, sect. 2, vol. 1 (Am. Physiological Soc., Bethesda 1979).

28 Beznak, M.; Hajdu, J.: The effect of extirpation of the stellage ganglion on mechanical heart hypertrophy of Albino rats. Arch. biol. hung. (II) 18: 11–18 (1948).

29 Bloom, G.; Östlund, E.; Euler, U.S. von; Lishajko, F.; Ritzén, M.; Adams-Ray, J.: Studies on catecholamine-containing granules of specific cells in cyclostome hearts. Acta physiol. scand. 53: suppl. 185, pp. 1–34 (1961).

30 Borchard, F.: The adrenergic nerves of the normal and the hypertrophied heart. Norm. pathol. Anat. 33: 1–68 (1978).

31 Braunwald, E.: Protection of the ischemic myocardium. Harvey Lect. 71: 247–282 (1978).

32 Braunwald, E.: Coronary spasm and acute myocardial infarction – new possibility for treatment and prevention. New Engl. J. Med. 299: 1301–1303 (1978).

33 Braunwald, E.: The role of catecholamines in heart failure; in Menezy, Caldwell, Catecholamines and the heart. Int. Congr. and Symp., series No. 8; pp. 31–45 (Royal Society of Medicine, London/Academic Press, London/Grune & Stratton, New York 1979).

34 Braunwald, E.; Harrison, D.C.; Chidsey, C.A.: The heart as an endocrine organ. Am. J. Med. 36: 1–4 (1964).

35 Braunwald, E.; Maroko, P.R.: Intra-aortic balloon counterpulsation: an assessment. Ann. intern. Med. 76: 659–691 (1972).

36 Braunwald, E.; Ross, J., Jr.; Sonnenblick, E.H.: Mechanisms of contraction of the normal and failing heart; 2nd ed., pp. 200–231 (Little, Brown, Boston 1976).

37 Braunwald, E.; Ross, J., Jr.; Sonnenblick, E.H.: Disorders of myocardial function; in Thorn, Adams, Braunwald, Isselbacher, Petersdorf, Harrison's principles of internal medicine, pp. 1167–1177 (McGraw-Hill, New York 1977).

38 Brewer, G.; Larson, P.S.; Schroeder, A.R.: On the effect of the epinephrine on blood potassium. Am. J. Physiol. *126:* 708–712 (1939).

39 Brooks, W.W.; Verrier, R.L.; Lown, B.: Extra-adrenergic mechanisms responsible for the effects of glucose-insulin-potassium solution on vulnerability to ventricular fibrillation. Am. J. Cardiol. *47:* 251–257 (1981).

40 Brown, A.M.: Motor innervation of the coronary arteries of the cat. J. Physiol., Lond. *198:* 311–328 (1968).

41 Brown, A.M.; Malliani, A.: Spinal reflexes initiated by coronary receptors. J. Physiol., Lond. *212:* 685–705 (1971).

42 Brown, B.C.: Coronary vasospasm: observations linking the clinical spectrum of ischemic heart disease to the dynamic pathology of coronary atherosclerosis. Archs intern. Med. *141:* 716–722 (1981).

43 Burch, G.E.; Phillips, J.H.: The large upright T wave as an electrocardiographic manifestation of intracranial disease. Sth. med. J., Nashville *61:* 331–336 (1968).

44 Burgess, A.M.; Waterman, G.W.; Cutts, F.B.: Adrenal sympathetic syndrome with unusual variations in cardiac rhythm. Archs intern. Med. *58:* 433–447 (1936).

45 Buu, N.T.; Kuchel, O.: A new method for the hydrolysis of conjugated catecholamines. J. Lab. clin. Med. *90:* 680–684 (1977).

46 Cahill, G.F.; Monteith, J.C.: Use of dibenamine and norepinephrine in the operative management of pheochromocytoma. New Engl. J. Med. *244:* 657–661 (1951).

47 Cannom, D.S.; Rider, A.K.; Stinson, E.B.; Harrison, D.C.: Electrophysiologic studies in the denervated transplanted human heart. II. Response to norepinephrine, isoproterenol and propranolol. Am. J. Cardiol. *36:* 859–866 (1975).

48 Cannon, P.: Some newer aspects of electrocardiography: study of 16 cases of pheochromocytoma. Ir. J. med. Sci. *359:* 499–511 (1955).

49 Cannon, P.; Sjöstrand, T.: ECG changes seen in cases of pheochromocytoma compared with changes experimentally evoked by adrenaline. Scand. J. clin. Lab. Invest. *4:* 266–267 (1952).

50 Cannon, W.B.: A law of denervation. Am. J. med. Sci. *198:* 737–750 (1939).

51 Cannon, W.B.; Lissak, K.: Evidence for adrenaline in adrenergic neurons. Am. J. Physiol. *125:* 765–777 (1939).

52 Cannon, W.B.; Rosenblueth, A.: Studies on conditions of activity in endocrine organs. XXIX. Sympathin E and sympathin I. Am. J. Physiol. *104:* 557–574 (1933).

53 Castelden, J.I.M.: The effect of adrenalin on the plasma potassium level. Clin. Sci. *3:* 241–245 (1938).

54 Chapman, C.B.; Jensen, D.; Wildenthal, K.: On circulatory control mechanisms in the Pacific hagfish. Circulation Res. *12:* 427–440 (1963).

55 Chatterjee, K.; Parmley, W.W.: The role of vasodilator therapy in heart failure. Prog. cardiovasc. Dis. *19:* 301–325 (1977).

56 Cheng, T.O.; Bashour, T.T.: Striking cardiographic changes associated with pheochromocytoma masquerading as ischemic heart disease. Chest *70:* 397–399 (1976).

57 Chiariello, M.; Brwetti, G.; Rosa, G. de; Acunzo, R.; Petillo, F.; Rengo, F.; Con-
    dorilli, M.: Protective effects of simultaneous alpha- and beta-adrenergic receptor
    blockade on myocardial cell necrosis after coronary arterial occlusion in rats. Am. J.
    Cardiol. 46: 249–254 (1980).
58 Chidsey, C.A.; Braunwald, E.; Morrow, A.G.: Catecholamine excretion and
    cardiac stores of norepinephrine in congestive heart failure. Am J. Med. 39: 442–451
    (1965).
59 Chidsey, C.A.; Braunwald, E.; Morrow, A.G.; Mason, D.T.: Myocardial norepi-
    nephrine concentration in man: effects of reserpine and congestive heart failure.
    New Engl. J. Med. 269: 653–658 (1963).
60 Chidsey, C.A.; Harrison, D.C.; Braunwald, E.: Augmentation of the plasma norepi-
    nephrine response to exercise in patients with congestive heart failure. New Engl. J.
    Med. 267: 650–654 (1962).
61 Chidsey, C.A.; Harrison, D.C.; Braunwald, E.: Release of norepinephrine from the
    heart by vasoactive amines. Proc. Soc. exp. Biol. Med. 109: 488–490 (1962).
62 Chidsey, C.A.; Kaiser, G.A.; Braunwald, E.: Biosynthesis of norepinephrine in the
    isolated canine heart. Science 139: 828–829 (1963).
63 Chidsey, C.A.; Kaiser, G.A.; Sonnenblick, E.H.; Spann, J.F.; Braunwald, E.:
    Cardiac norepinephrine stores in experimental heart failure in the dog. J. clin. Invest.
    43: 2386–2393 (1964).
64 Chidsey, C.A.; Sonnenblick, E.H.; Morrow, A.G.; Braunwald, E.: Norepinephrine
    stores and contractile force of papillary muscle from the failing human heart.
    Circulation 33: 43–51 (1966).
65 Christensen, N.J.: A sensitive assay for the determination of dopamine in plasma.
    Scand. J. clin. Lab. Invest. 31: 343–346 (1973).
66 Cohen, J.: Role of endocrine factors in the pathogenesis of cardiac hypertrophy.
    Circulation Res. 35: suppl. II, pp. 49–57 (1974).
67 Cohn, J.N.; Franciosa, J.A.: Vasodilator therapy of cardiac failure. 1. New Engl.
    J. Med. 297: 27–31 (1977).
68 Cohn J.N.; Franciosa, J.A.: Vasodilator therapy of cardiac failure. 2. New Engl.
    J. Med. 297: 254–258 (1977).
69 Cohn, J.N.; Taylor, N.; Vrobel, T.; Moskowitz, R.: Contrasting effects of vaso-
    dilators on heart rate and plasma catecholamines in patients with hypertension and
    heart failure (Abstract). Clin. Res. 26: 547A (1978).
70 Conway, J.: A vascular abnormality in hypertension. A study of bloodflow in the
    forearm. Circulation 27: 520–529 (1963).
71 Cooper, T.; Gilbert, J.W., Jr.; Bloodwell, R.D.; Crout, J.R.: Chronic extrinsic
    cardiac denervation by regional neural ablation: Description of the operation,
    verification of the denervation and it's effects on myocardial catecholamines. Circula-
    tion Res. 9: 275–281 (1961).
72 Cooper, T.; Willman, V.L.; Hanlon, C.R.: Drug responses of the transplanted
    heart. Dis. Chest 45: 284–287 (1964).
73 Cooper, T.; Willman, V.L.; Jellinek, M.; Hanlon, C.R.: Heart autotransplantation:
    effect on myocardial catecholamines and histamine. Science 138: 40–41 (1962).
74 Corbett, J.L.; Kerr, J.H.; Prys-Roberts, C.; Smith, A.C.; Spalding, J.M.K.: Cardio-
    vascular disturbances in severe tetanus due to over-activity of the sympathetic
    nervous system. Anesthesia 24: 198–212 (1969).

75 Corr, P.B.; Gillis, R.A.: Effect of autonomic neural influences in the cardiovascular changes induced by coronary occlusion. Am. Heart J. *89:* 766–774 (1975).

76 Coulson, R.L.; Yazdanfar, S.; Rubio, E.; Bove, A.A.; Lemole, G.M.; Spann, J.F., Jr.: Recuperative potential of cardiac muscle following relief of pressure overload hypertrophy and right ventricular failure in the cat. Circulation Res. *40:* 41–49 (1977).

77 Covell, J.W.; Chidsey, C.A.; Braunwald, E.: Reduction of the cardiac response to postganglionic sympathetic nerve stimulation in experimental heart failure. Circulation Res. *19:* 51–56 (1966).

78 Cox, W.F.; Robertson, H.F.: The effect of stellate ganglionectomy on the cardiac function of intact dogs. Am. Heart J. *12:* 285–300 (1936).

79 Creese, I.; Sibley, D.R.; Leff, S.; Hamblin, M.: Dopamine receptors: subtypes, localization and regulation. Fed. Proc. *40:* 147–152 (1981).

80 Dahlström, A.; Fuxe, K.; Mya-Tu, M.; Zetterström, B.E.M.: Observations of adrenergic innervation of the dog heart. Am. J. Physiol. *209:* 689–692 (1965).

81 De Champlain, J.: Experimental aspects of the relationships between the autonomic nervous system and catecholamines in hypertension; in Genest, Koiw, Kuchel, Hypertension, pp. 76–92 (McGraw-Hill, New York 1977).

82 De Champlain, J.; Farley, L.; Cousineau, D.; Amerigen, M.-R. van: Circulating catecholamine levels in human and experimental hypertension. Circulation Res. *38:* 109–114 (1976).

83 De Champlain, J.; Krakoff, L.R.; Axelrod, J.: Relationship between sodium intake and norepinephrine storage during the development of experimental hypertension. Circulation Res. *23:* 479–491 (1968).

84 De Champlain, J.; Krakoff, L.; Axelrod, J.: Interrelationships of sodium intake, hypertension, and norepinephrine storage in the rat. Circulation Res. *24:* 75–92 (1969).

85 Dempsey, P.J.; Cooper, T.: Supersensitivity of the chronically denervated feline heart. Am. J. Physiol. *215:* 1245–1249 (1968).

86 DeQuattro, V.; Nagatsu, T.; Mendez, A.; Verska, J.: Determinants of cardiac noradrenaline depletion in human congestive failure. Cardiovasc. Res. *7:* 344–350 (1973).

87 Dhalla, N.S.; Naidu, K.J.; Bhagat, B.: Biochemical basis of heart function. Relation of catecholamine stores and contractile force in an isolated rat heart. Cardiovasc. Res. *5:* 376–382 (1971).

88 Dollery, C.T.: The physiological role of catecholamine neurotransmitters in the heart; in Mezey, Caldwell, Catecholamines and the heart. Int. Congr. Symp. Ser. No. 8, pp. 13–16 (Royal Society of Medicine, London/Academic Press, London/ Grune & Stratton, New York 1979).

89 Donald, D.E.: Brief reviews – Myocardial performance after excision of the extrinsic cardiac nerves in the dog. Circulation Res. *34:* 417–424 (1974).

90 Donald, D.E.; Ferguson, D.A.; Milburn, S.E.: Effect of beta-adrenergic receptor blockade on racing performance of greyhounds with normal and with denervated hearts. Circulation Res. *22:* 127–134 (1968).

91 Donald D.E.; Milburn, S.E.; Shepherd, J.T.: Effect of cardiac denervation on the maximal capacity for exercise in the racing greyhound. J. appl. Physiol. *19:* 849–852 (1964).

92 Donald, D.E.; Samueloff, S.L.: Exercise tachycardia not due to blood-borne agents in canine cardiac denervation. Am. J. Physiol. *211:* 703–711 (1966).

93 Donald, D.E.; Shepherd, J.T.: Response to exercise in dogs with cardiac denervation· Am. J. Physiol. *205:* 393–400 (1963).

94 Douglas, W.W.: Stimulus-secretion coupling: the concept and clues from chromaffin and other cells. Br. J. Pharmacol. *34:* 451–474 (1968).

95 Doyle, A.E.: Endogenous catecholamine content of cardiac muscle in sodium-loaded and sodium-depleted rats. Lancet *i:* 1399–1400 (1968).

96 Durant, J.; Soloff, L.A.: Arrhythmic crisis of pheochromocytoma. Lancet *ii:* 124–126 (1962).

97 Dury, A.: The effect of epinephrine and insulin on the serum potassium level in man. Endocrinology *49:* 663–670 (1951).

98 Dustan, H.P.; Tarazi, R.C.; Frohlich, E.D.: Functional correlates of plasma renin activity in hypertensive patients. Circulation *41:* 555–567 (1970).

99 Ebert, P.: Personal communication cited by Schall, Wallace, Sealy. Cardiovasc. Res. *3:* 241–244 (1969).

100 Ebert, P.A.; Allgood, R.J.; Sabiston, D.C., Jr.: Antiarrhythmic effects on cardiac denervation. Ann. Surg. *168:* 728–735 (1968).

101 Ebert, P.A.; Vanderbeek, R.B.; Allgood, R.J.; Sabiston, D.C.: Effect of chronic cardiac denervation on arrhythmias after coronary artery ligation. Cardiovasc. Res. *4:* 141–147 (1970).

102 Eckberg, D.I.; Drabinsky, M.; Braunwald, E.: Defective cardiac parasympathetic control in patients with heart disease. New Engl. J. Med. *285:* 877–883 (1971).

103 Ehinger, B.; Falck, B.; Sporrong, B.: Adrenergic fibers to the heart and to peripheral vessels. Symp. Elec. Activ. Innerv. Blood Vessels, Cambridge 1966. Biblthca anat., No. 8, pp. 35–45 (Karger, Basel 1966).

104 Eich, R.H.; Peters, R.J.; Cuddy, R.P.; Smulyan, H.; Lyons, R.H.: The hemodynamics in labile hypertension. Am. Heart J. *63:* 188–195 (1962).

105 Ellrodt, G.; Chew, C.Y.C.; Singh, B.N.: Therapeutic implications of slow-channel blockade in cardiocirculatory disorders. Circulation *62:* 669–679 (1980).

106 Engelman, K.; Sjoerdsma, A.: Chronic medical therapy for pheochromocytoma: a report of four cases. Ann. intern. Med. *61:* 229–241 (1964).

107 Epstein, S.E.; Braunwald, E.: The effect of beta-adrenergic blockade on patterns of urinary sodium excretion: studies in normal subjects and in patients with heart disease. Ann. intern. Med. *65:* 20–27 (1966).

108 Epstein, S.E.; Robinson, B.F.; Kahler, R.L.; Braunwald, E.: Effects of beta-adrenergic blockade on the cardiac response to maximal and submaximal exercise in man. J. clin. Invest. *44:* 1745–1753 (1965).

109 Erickson, H.H.; Stone, H.L.: Cardiac beta-adrenergic receptors and coronary haemodynamics in the conscious dog during hypoxic hypoxia. Aerospace Med. *43:* 422–428 (1972).

110 Esler, M.D.; Julius, S.; Randall, O.S.; Ellis, C.N.; Kashima, T.: Relation of renin status in neurogenic vascular resistance in borderline hypertension. Am. J. Cardiol. *36:* 708–715 (1975).

111 Esler, M.D.; Nestel, P.J.: Sympathetic responsiveness to head-up tilt in essential hypertension. Clin. Sci. *44:* 213–226 (1973).

112 Esperson, T.; Dahl-Iversen, E.: The clinical picture and treatment of pheochromocytomas of the suprarenal. Acta chir. scand. *94:* 271–290 (1946).

113 Euler, U.S. von: A specific sympathomimetic ergone in adrenergic nerve fibers

(sympathin) and its relation to adrenaline and noradrenaline. Acta physiol. scand. *12:* 73–97 (1946).

114   Euler, U.S. von: Noradrenaline (Thomas, Springfield 1956).

115   Euler, U.S. von; Franksson, C.; Hellström, J.: Adrenaline and noradrenalin output in urine after unilateral and bilateral adrenalectomy in man. Acta physiol. scand. *31:* 1–5 (1954).

116   Euler, U.S. von; Ström, G.: Present status of diagnosis and treatment of pheochromocytoma. Circulation *15:* 5–13 (1957).

117   Evarts, E.V.; Gillespie, L.; Fleming, T.C.; Sjoerdsma, A.: Relative lack of pharmacologic action of 3-methoxy analog of norepinephrine. Proc. Soc. exp. Biol. Med. *98:* 74–76 (1958).

118   Feigl, E.O.: Sympathetic control of coronary circulation. Circulation Res. *20:* 262–271 (1967).

119   Feigl, E.O.: Control of myocardial oxygen tension by sympathetic coronary vasoconstriction in the dog. Circulation Res. *37:* 88–95 (1975).

120   Feigl, E.O.: Reflex parasympathetic coronary vasodilation elicited from cardiac receptors in the dog. Circulation Res. *37:* 175–182 (1975).

121   Fernandes, M.; Onesti, G.; Fiorentini, R.; Kim, K.E.; Swartz, C.: Effect on chronic administration of propranolol on the blood pressure and heart weight in experimental renal hypertension. Life Sci. *18:* 967–970 (1976).

122   Fischer, J.E.; Horst, W.D.; Kopin, I.J.: Norepinephrine metabolism in hypertrophied rat hearts. Nature, Lond. *207:* 951–953 (1965).

123   Fölkow, B.: The haemodynamic consequences of adaptive structural changes of the resistance vessels in hypertension. Clin. Sci. mol. Med. *41:* 1–12 (1971).

124   Forde, T.P.; Yormak, S.S.; Killip, T., III: Reflex bradycardia and nodal escape rhythm in pheochromocytoma. Am. Heart J. *76:* 388–392 (1968).

125   Fowlis, R.A.F.; Sang, C.T.M.; Lundy, P.M.; Ahula, S.P.; Colhoun, H.: Experimental coronary artery ligation in conscious dogs six months after bilateral cardiac sympathectomy. Am. Heart J. *88:* 748–757 (1974).

126   Franco-Morcelli, R.; Baudouin-Legros, M.; Mendonça, M. de; Guicheney, P.; Meyer, P.: Plasma catecholamines in essential human hypertension and in DOCA-salt hypertension in the rat; in Birkenhager, Falke, Circulating catecholamines and blood pressure, pp. 27–38 (Bunge, Utrecht 1978).

127   Frankel, H.L.; Mathias, C.J.: Cardiovascular aspects of autonomic dysreflexia since Guttmann and Whitteridge. Paraplegia *17:* 46–51 (1979).

128   French, C.; Campagna, F.A.: Pheochromocytoma with shock, marked leukocytosis, and unusual electrocardiograms. Case report and review of the literature. Ann. intern. Med. *55:* 127–134 (1961).

129   Frohlich, E.D.; Dustan, H.P.; Page, I.H.: Hyperdynamic beta-adrenergic circulatory state. Archs intern. Med. *117:* 614–619 (1966).

130   Frohlich, E.D.; Kozul, V.J.; Tarazi, R.C.; Dustan, H.P.: Physiological comparison of labile and essential hypertension. Circulation Res. *27:* suppl. I, pp. 55–63 (1970).

131   Frohlich, E.D.; Tarazi, R.D.; Dustan, H.P.: Hyperdynamic beta-adrenergic circulatory state. Archs intern. Med. *123:* 1–7 (1969).

132   Furchgott, R.F.: The classification of adrenoceptors (adrenergic receptors). An evaluation from the standpoint of receptor theory; in Blaschko, Muscholl, Catecholamines, pp. 283–335 (Springer, Berlin 1972).

133 Futterweit, W.; Allen, L.; Moser, M.: Pheochromocytoma with radioactive iodine uptake and eletrocardiographic abnormalities. Metabolism *11:* 589–599 (1962).

134 Gaffney, T.E.; Braunwald, E.: Importance of the adrenergic nervous system in the support of circulatory function in patients with congestive heart failure. Am. J. Med. *34:* 320–324 (1963).

135 Gaffney, T.E.; Braunwald, E.; Cooper, T.: Analysis of the acute circulatory effects of quanethidine and bretylium. Circulation Res. *10:* 83–88 (1962).

136 Gaffney, T.E.; Morrow, D.H.; Chidsey, C.A.: The role of myocardial catecholamines in the response to tyramine. J. Pharmac. exp. Ther. *137:* 301–305 (1962).

137 Gans, J.H.; Cater, M.R.: Norepinephrine induced cardiac hypertrophy in dogs. Life Sci. *9:* 731–734 (1970).

138 Gauthier, P.; Nadeau, R.; deChamplain, J.: The development of sympathetic innervation and the functional state of the cardiovascular system in newborn dogs. Can. J. Physiol. Pharmacol. *53:* 763–776 (1975).

139 Gazes, P.G.; Richardson, J.A.; Woods, E.F.: Plasma catecholamine concentrations in myocardial infarction and angina pectoris. Circulation *19:* 657–661 (1959).

140 Gillis, R.A.; Corr, P.B.; Pace, D.G.; Evans, D.E.; DiMicco, J.; Pearle, D.L.: Role of the nervous system in experimentally induced arrhythmias. Cardiology *61:* 37–49 (1976).

141 Gillis, D.; Pearle, L.; Hoekman, T.: Failure of β-adrenergic receptor blockade to prevent arrhythmias induced by sympathetic nerve stimulation. Science *185:* 70–72 (1974).

142 Giotti, A.; Ledda, F.; Mannaioni, P.F.: Effects of noradrenaline and isoprenaline in combination with alpha- and beta-receptor blocking substances on the action potential of cardiac Purkinje fibers. J. Physiol., Lond. *229:* 99–113 (1973).

143 Glaubiger, G.; Lefkowitz, R.J.: Elevated beta-adrenergic receptor number after chronic propranolol treatment. Biochem. biophys. Res. Commun. *78:* 720–725 (1977).

144 Gold, H.K.; Leinbach, R.C.; Maroko, P.R.: Reduction of myocardial injury in patients with acute infarction by propranolol (Abstract). Circulation *50:* suppl. III, p. 33 (1974).

145 Goldberg, H.C.; Marsden, C.A.: Catechol-*O*-methyl transferase: pharmacological aspects and physiological role. Pharmacol. Rev. *27:* 135–206 (1975).

146 Goldberg, L.I.: Cardiovascular and renal actions of dopamine: potential clinical applications. Pharmacol. Rev. *24:* 1–29 (1972).

147 Goldberg, L.I.: Dopamine – clinical uses of an endogenous catecholamine. New Engl. J. Med. *291:* 707–710 (1974).

148 Goldberg, L.I.; Kohli, J.D.; Kotake, A.N.; Volkman, P.A.: Characteristics of the vascular dopamine receptor: comparison with other receptors. Fed. Proc. *37:* 2396–2402 (1978).

149 Goldberg, L.I.; Volkman, P.H.; Kohli, J.D.: A comparison of the vascular dopamine receptor with other dopamine receptors. Annu. Rev. Pharmacol. Toxicol. *18:* 57–79 (1978).

150 Goldenberg, M.; Rapport, M.M.: Nor-epinephrine and epinephrine in human urine (Addison's disease, essential hypertension, pheocromocytoma). J. clin. Invest. *30:* 641–642 (1951).

151 Goldstein, R.A.; Passamani, E.R.; Roberts, R.: A comparison of digoxin and dobutamine in patients with acute infarction and cardiac failure. New Engl. J. Med. *303:* 846–850 (1980).

152 Goldstein, R.E.; Beiser, G.D.; Stampfer, M.; Epstein, S.E.: Impairment of autonomically mediated heart rate control in patients with cardiac dysfunction. Circulation Res. *36:* 571–578 (1975).

153 Goodall, McC.: The presence of noradrenaline and an unknown sympathetic factor in cattle heart. Acta physiol. scand. *20:* 137–152 (1950).

154 Goodall, McC.; Kirshner, N.: Effect of cervico-thoracic ganlonectomy on the adrenaline and noradrenaline content of the mammalian heart. i. clin. Invest. *35:* 649–656 (1956).

155 Granata, L.; Olsson, R.A.; Huvos, A.; Gregg, D.E.: Coronary inflow of oxygen usage following cardiac sympathetic nerve stimulation in unanesthetized dogs. Circulation Res. *16:* 114–120 (1965).

156 Green, K.G.; Chamberlain, D.A.; Fulton, R.M.; Hamer, N.A.J.; Oliver, M.F.; Pentecost, B.L.: Improvement in prognosis of myocardial infarction by long-term beta-adrenoceptor blockade using practolol: a multicentre international study. Br. med. J. *iii:* 735–740 (1975).

157 Gregg, D.E.; Khouri, E.M.; Donald, D.E.; Lowensohn, H.S.; Stanislaw, P.: Coronary circulation in the conscious dog with cardiac neural ablation. Circulation Res. *31:* 129–144 (1972).

158 Grondin, C.M.; Limet, R.: Sympathetic denervation in association with coronary artery grafting in patients with Prinzmetal's angina. Ann. thor. Surg. *23:* 111–117 (1977).

159 Grovier, W.C.; Mosal, N.C.; Whittington, P.; Broom, A.H.: Myocardial alpha and beta adrenergic receptors as demonstrated by atrial functional refractory-period changes. J. Pharmac. exp. Ther. *154:* 255–263 (1966).

160 Guyton, R.A.; Bianco, J.A.; Ostheimer, G.W.; Shanohan, E.H.; Daggett, W.M.: Adrenergic control of ventricular performance in normal and cardiac denervated dogs. Am. J. Physiol. *223:* 1021–1028 (1972).

161 Guyton, A.C.; Coleman, T.G.: Quantitative analysis of the pathophysiology of hypertension. Circulation Res. *24:* suppl. I, pp. 1–19 (1969).

162 Hageman, G.R.; Geis, W.P.; Kaye, M.P.: Regional myocardial norepinephrine in the functionally reinnervated heart. Fed. Proc. *32:* 344 (1973).

163 Hageman, G.R.; Goldberg, J.M.; Armour, J.A.; Randall, W.C.: Cardiac dysrhythmias induced by autonomic nerve stimulation. Am. J. Cardiol. *32:* 822–830 (1973).

164 Hamilton, M.; Litchfield, J.W.; Peart, W.S.; Sowry, G.S.C.: Pheochromocytoma. Br. Heart J. *15:* 241–249 (1953).

165 Han, J.; De Jalon, G.; Moe, G.K.: Adrenergic effects on ventricular vulnerability. Circulation Res. *14:* 516–524 (1964).

166 Han, J.; Moe, C.K.: Nonuniform recovery of excitability in ventricular muscle. Circulation Res. *14:* 44–60 (1964).

167 Harris, A.S.: Genesis of ventricular tachycardia and fibrillation following coronary occlusion; in Dreifus, Likoff, Mechanisms and therapy of cardiac arrhythmias, pp. 345–378 (Grune & Stratton, New York 1964).

168 Harris, A.S.; Estandia, A.; Tillotson, R.F.: Ventricular ectopic rhythms and ventricular fibrillation following cardiac sympathectomy and coronary occlusion. Am. J. Physiol. *165:* 505–512 (1951).

169 Harrison, D.C.; Chidsey, C.A.; Goldman, R.; Braunwald, E.: Relationships between the release and tissue depletion from the heart by guanethidine and reserpine. Circulation Res. *10:* 256–263 (1963).

170 Harrison, D.C.; Mason, J.W.: Effects of catecholamines and adrenergic innervation on cardiac conduction and arrhythmias; in Mezey, Caldwell, Catecholamines and the heart. Int. Congr. Symp. ser., No.8, pp. 17–30 (Royal Society of Medicine, London/Academic Press, London/Grune & Stratton, New York 1979).

171 Hegglin, R.; Holzmann, M.: Elektrokardiographische Befunde beim Paragangliom der Nebenniere. Dt. Arch. klin. Med. *180:* 681–691 (1937).

172 Henning, M.: Noradrenaline turnover in renal hypertensive rats. J. Pharm. Pharmac. *21:* 61–63 (1969).

173 Higgins, C.B.; Vatner, S.F.; Braunwald, E.: Regional hemodynamic effects of a digitalis glycoside in the conscious dog with and without experimental heart failure. Circulation Res. *30:* 406–417 (1972).

174 Higgins, C.B.; Vatner, S.F.; Braunwald, E.: Parasympathetic control of the heart. Pharmacol. Rev. *25:* 119–155 (1973).

175 Higgins, C.B.; Vatner, S.F.; Eckberg, D.L.; Braunwald, E.: Alterations in the baro-receptor reflex in conscious dogs with heart failure. J. clin. Invest. *51:* 715–724 (1972).

176 Higgins, C.B.; Vatner, S.F.; Franklin, D.; Braunwald, E.: Effect of experimentally produced heart failure on the peripheral vascular response to severe exercise in conscious dogs. Circulation Res. *31:* 186–194 (1972).

177 Hillarp, N.A.: The construction and functional organization of the autonomic inner-vation. Acta physiol. scand. *46:* suppl. 157, pp. 1–38 (1959).

178 Hillis, L.D.; Braunwald, E.: Coronary-artery spasm. New Engl. J. Med. *299:* 695–702 (1978).

179 Hingerty, D.; O'Boyle, A.: Clinical chemistry of the adrenal medulla, pp. 78–88 (Thomas, Springfield 1972).

180 Hoffman, B.F.: Appraisal of effects of catecholamines on cardiac electrical activity. Ann. N.Y. Acad. Sci. *139:* 914–939 (1967).

181 Hoffman, B.F.: Neural influences on cardiac electrical excitability and rhythm; in Randall, Neural regulation of the heart, pp. 291–312 (Oxford University Press, New York 1977).

182 Hoffman, B.F.; Bigger, J.T., Jr.: Antiarrhythmic drugs; in DiPalma, Drill's pharma-cology in medicine; 4th ed., pp. 824–852 (McGraw-Hill, New York 1971).

183 Hoffmann, F.; Hoffmann, E.J.; Middleton, S.; Talisnik, J.: The stimulation effect of acetycholine on the mammalian heart and the liberation of an epinephrine-like substance by the isolated heart. Am. J. Physiol. *144:* 189–198 (1945).

184 Hökfelt, B.: Noradrenaline and adrenaline in mammalian tissues; distribution under normal and pathological conditions with special reference to the endocrine system. Acta physiol. scand. *25:* suppl. 92, pp. 1–134 (1951).

185 Hollenberg, M.; Carriere, S.; Barger, A.C.: Biphasic action of acetylcholine on ven-tricular myocardium. Circulation Res. *16:* 527–536 (1965).

186 Hurst, J.W.; Logue, R.B.; Walter, P.F.: The clinical recognition and medical management of coronary artherosclerotic heart disease; in Hurst, Logue, Schlant, Wenger, The heart, pp. 1156–1290 (McGraw-Hill, New York 1978).

187 Imai, S.; Shigei, T.; Hashimoto, K.: Cardiac actions of methoxamine with special reference to its antagonistic action to epinephrine. Circulation Res. *9:* 552–560 (1961).

188 Iriuchijima, J.; Kumada, M.: Efferent cardiac vagal discharge of the dog in response to electrical stimulation of sensory nerves. Jap. J. Physiol. *13:* 599–605 (1963).

189 Iseri, L.T.; Henderson, H.W.; Derr, J.W.: Use of adrenolytic drug, Regitine, in pheochromocytoma. Am. Heart J. *42:* 129–136 (1951).

190 Iversen, L.L.: Uptake and storage of norepinephrine in sympathetic nerves (Cambridge University Press, London, 1967).

191 Iversen, L.L.: Uptake of circulating catecholamines into tissues; in Blaschko, Sayers, Smith, Handbook of physiology, vol. VI., pp. 713–722 (Am. Physiological Society, Washington 1975).

192 Iversen, L.L.: Dopamine receptors in the brain: a dopamine-sensitive adenylate cyclase models synaptic receptors, illuminating antipsychotic drug action. Science *188:* 1084–1089 (1975).

193 Jacobson, W.E.; Hammarsten, J.F.; Heller, B.I.: The effects of adrenaline upon renal function and electrolyte excretion. J. clin. Invest. *30:* 1503–1506 (1951).

194 James, T.N.: The chronotropic action of ATP and related compounds studied by direct fusion of the sinus node. J. Pharmac. exp. Ther. *149:* 233–247 (1965).

195 James, T.N.; Bear, E.S.; Lang, K.F.; Green, E.W.: Evidence for adrenergic alpha receptor depressant activity in the heart. Am. J. Physiol. *215:* 1366–1375 (1968).

196 James, T.N.; Bear, E.S.; Lang, K.F.; Green, E.W.; Winkler, H.H.: Adrenergic mechanism in the sinus node. Archs intern. Med. *125:* 512–547 (1970).

197 James, T.N.; Isobe, J.H.; Urthaler, F.: Analysis of components in a cardiogenic hypertensive chemoreflex. Circulation *52:* 179–192 (1975).

198 James, T.N.; Spence, C.A.: Cholinesterase within the sinus node and AV node of the human heart. Anat. Rec. *155:* 151–161 (1966).

199 Jewitt, D.E.; Mercer, C.J.; Reid, D.; Valori, C.; Thomas, M.; Shillingford, J.P.: Free noradrenaline and adrenaline excretion in relation to the development of cardiac arrhythmias and heart failure in patients with acute myocardial infarction. Lancet *i:* 635 (1969).

200 Juhász-Nagy, A.; Szentiványi, M.: Effect of alpha-adrenergic coronary constriction on myocardial tissue blood flow. Archs Int. Pharmacodyn. Thér. *206:* 19–20 (1973).

201 Julius, S.; Conway, J.: Hemodynamic studies in patients with borderline blood pressure elevation. Circulation *38:* 282–288 (1968).

202 Julius, S.; Esler, M.: Autonomic nervous and cardiovascular regulation in borderline hypertension. Am. J. Cardiol. *36:* 685–696 (1975).

203 Julius, S.; Esler, M.; Randall, O.: Neurogenic borderline hypertension (Abstract). Circulation *50:* suppl. III, p. 107 (1974).

204 Julius, S.; Esler, M.D.; Randall, O.S.: Role of the autonomic nervous system in mild human hypertension. Clin. Sci. Mol. Med. *48:* suppl., pp. 243s–252s (1975).

205 Karlsberg, R.P.; Cryor, P.E.; Roberts, R.: Serial plasma catecholamine response early in the course of clinical acute myocardial infarction: relationship to infarct extent and mortality. Am. Heart J. *102:* 24–29 (1981).

206 Katz, A.M.: Physiology of the heart (Raven Press, New York 1977).

207 Kaufman, S.; Friedman, S.: Dopamine-β-hydroxylase. Pharmacol. Rev. *17:* 71–100 (1965).

208 Kaumann, A.J.: Potentiation of the effects of isoprenaline and noradrenaline by hydrocortisone in cat heart muscle. Arch. Pharmacol. *273:* 134–153 (1972).

209 Kaye, M.P.: Denervation and reinnervation of the heart; in Randall, Neural regulation of the heart, pp. 347–378 (Oxford University Press, New York 1977).

210 Kaye, M.P.; Witzke, D.J.; Wells, D.J.; Fuster, V.: Nerve growth factor-induced hypertrophic cardiomyopathy in dogs (unpublished).

211 Kelliher, G.J.; Roberts, J.: A study of the antiarrhythmic action of beta blocking agents. Am. Heart J. *87:* 458–467 (1974).

212 Kent, K.M.; Epstein, S.E.; Cooper, T.; Jacobowitz, D.C.: Cholinergic innervation of the canine and human ventricular conducting system, anatomic and electro-physiologic correlations. Circulation *50:* 948–955 (1974).

213 Keys, A.: The response of the plasma potassium level in man to the administration of epinephrine. Am. J. Physiol. *121:* 325–330 (1938).

214 Kezdi, P.; Kordenat, R.K.; Misra, S.N.: Reflex inhibitory effects of vagal afferents in experimental myocardial infarction. Am. J. Cardiol. *33:* 853–860 (1974).

215 Khan, M.I.; Hamilton, J.T.; Manning, G.W.: Protective effect of beta adrenoceptor blockade in experimental coronary occlusion in conscious dogs. Am. J. Cardiol. *30:* 832–837 (1972).

216 Kimata, S.: Distribution of cardiac sympathetic nerves. Jap. Circul. J. *29:* 17–20 (1965).

217 Kliks, B.R.; Burgess, M.J.; Abildskov, J.A.: Influence of sympathetic tone on ventricular fibrillation threshold during experimental coronary occlusion. Am. J. Cardiol. *36:* 45–49 (1975).

218 Kline, I.K.: Myocardial alterations associated with pheochromocytoma. Am. J. Path. *38:* 539–552 (1961).

219 Kopin, I.J.: Storage and metabolism of catecholamines: the role of monoamine oxidase. Pharmacol. Rev. *16:* 179–191 (1964).

220 Kositzky, G.I.: Regulation function of the intercardiac nervous system (Abstract). Proc. Int. Un. Physiol. Sci. *9:* 319 (1971).

221 Krakoff, L.R.; Buccino, R.A.; Spann, J.F., Jr.; De Champlain, J.: Cardiac catechol-*O*-methyltransferase and monoamine oxidase activity in congestive heart failure. Am. J. Physiol. *215:* 549–552 (1968).

222 Krakoff, L.R.; De Champlain, J.; Axelrod, J.: Abnormal storage of norepinephrine in experimental hypertension in the rat. Circulation, Res. *21:* 583–591 (1967).

223 Kramer, R.S.; Mason, D.T.; Braunwald, E.: Augmented sympathetic neurotrans-mitter stores in peripheral vascular bed of patients with congestive heart failure and cardiac norepinephrine depletion. Circulation *38:* 629–634 (1968).

224 Kuchel, O.; Buu, N.T.; Fountaine, A.; Hamet, P.; Unger, T.; Genest, J.: Free and conjugated plasma catecholamines in the venous effluent of various organs of hypertensive patients. Eur. J. clin. Invest. *7:* 75–76 (1977).

225 Kuchel, O.; Cuche, J.L.; Hamet, P.; Boucher, R.; Barbeau, A.; Genest, J.: The relationship between adrenergic nervous system and renin in labile hyperkinetic hypertension; in Genest, Koiw, Hypertension; pp. 118–125 (Springer, New York 1972).

226 LaBrosse, E.H.; Axelrod, J.; Kety, S.S.: *O*-Methylation: the principle route of metabolism of epinephrine in man. Science *128:* 593–594 (1958).

227 LaBrosse, E.H.; Hertting, G.: Bilary excretion of *dl*-epinephrine metabolites. Fed. Proc. *19:* 398–404 (1964).

228 Laks, M.M.; Morady, F.; Swan, H.J.C.; Myocardial hypertrophy produced by chronic infusion of subhypertensive doses of norepinephrine in the dog. Chest *64:* 75–78 (1973).

229 Langer, S.Z.: Presynaptic regulation of catecholamine release. Biochem. Pharmacol. *23:* 1793–1800 (1974).

230 Lavallée, M.; De Champlain, J.; Nadeau, R.A.; Yamaguchi, N.: Muscarinic inhibition of endogenous myocardial catecholamine liberation in the dog. Can. J. Physiol. Pharmacol. *56:* 642–649 (1978).

231 Lefer, L.G.; Ayers, C.R.: Norepinephrine metabolism in dogs with chronic renovascular hypertension. Proc. Soc. exp. Biol. Med. *132:* 278–280 (1969).

232 Lefkowitz, R.J.: β-adrenergic receptors; recognition and regulation. New Engl. J. Med. *295:* 323–328 (1976).

233 Lefkowitz, R.J.; Wessels, M.R.; Stadel, J.M.: Hormones, receptors, and cyclic AMP. Curr. Top. cell. Regul. *17:* 205–229 (1980).

234 Leinbach, R.C.; Gold, H.K.; Buckley, M.J.; Austen, W.G.; Sanders, C.A.: Reduction of myocardial injury during acute infarction by early application of intra-aortic balloon pumping and propranolol (Abstract). Circulation *48:* suppl. IV, p. 100 (1973).

235 Lemberg, L.; Catellanos, A., Jr.; Arcebal, A.G.: The use of propranolol in arrhythmias complicating acute myocardial infarction. Am. Heart J. *80:* 479–487 (1970).

236 Levitt, M.; Spector, S.; Sjoerdsma, A.; Udenfriend, S.: Elucidation of the rate-determining step in norepinephrine biosynthesis in the perfused guinea pig heart. J. Pharmac. exp. Ther. *148:* 1–8 (1965).

237 Levy, M.N.: Parasympathetic control of the heart; in Randall, Neural regulation of the heart, pp. 97–129 (Oxford University Press, New York 1977).

238 Levy, M.N.: Neural control of the heart: sympathetic vagal interactions; in Baan, Noordegraaf, Raines, Cardiovascular system dynamics, pp. 365–370 (MIT Press, Cambridge 1978).

239 Levy, M.N.; Blattberg, B.: Correlation of the mechanical responses of the heart with the norepinephrine overflow during cardiac sympathetic neuronal stimulation in the dog. Cardiovasc. Res. *11:* 481–488 (1977).

240 Liard, J.-F.; Tarazi, R.C.; Ferrario, C.M.; Manger, W.M.: Hemodynamic and humoral characteristics of hypertension induced by prolonged stellate ganglion stimulation in conscious dogs. Circulation Res. *36:* 455–464 (1975).

241 Libby, P.; Maroko, P.R.; Covell, J.W.; Malloch, C.I.; Ross, J., Jr.; Braunwald, E.: The effects of practolol on the extent of myocardial ischemic injury following experimental coronary occlusion and its effects on ventricular function in the normal and ischemic heart. Cardiovasc. Res. *7:* 167–173 (1973).

242 Lloyd-Mostyn, R.; Watkins, P.J.; Oram, S.: Autonomic neuropathy and the heart in diabetes. Br. Heart J. *36:* 397 (1974).

243 Loewi, O.: Über humorale Übertragbarkeit der Herznervenwirkung. I. Mitteilung. Pflügner's Arch. ges. Physiol. *189:* 239–242 (1921).

244 Louis, W.J.; Krauss, K.R.; Kopin, I.J.; Sjoerdsma, A.: Catecholamine metabolism in hypertensive rats. Circulation Res. *27:* 589–594 (1970).

245 Louis, W.J.; Spector, S.; Tabei, R.; Sjoerdsma, A.: Synthesis and turnover of norepinephrine in the heart of the spontaneously hypertensive rat. Circulation Res. *24:* 85–91 (1969).

246 Lown, B.; Verrier, R.L.: Neural activity and ventricular fibrillation. New Engl. J. Med. *294:* 1165–1170 (1976).

247 Lown, B.; Verrier, R.L.; Corbalan, R.: Psychologic stress and threshold for repetitive ventricular response. Science *182:* 834–836 (1973).

248 Luft, F.; Bloch, R.; Grim, C.; Henry, D.; Weinberger, M.: Sympathetic nervous system activity (SNSA) and salt balance. J. clin. Res. *26:* 365A (1978).

249 Lund-Johansen, P.: Hemodynamics in early essential hypertension. Acta med. scand. *183:* suppl. 482, pp. 1–101 (1968).

250 Machado, C.R.S.; Machado, A.B.M.; Chiari, C.A.: Recovery from heart norepinephrine depletion in experimental Chagas' disease. Am. J. trop. Med. Hyg. *27:* 20–24 (1978).

251 MacKeith, R.: Adrenal-sympathetic syndrome: chromaffin tissue tumor with paroxysmal hypertension. Br. Heart J. *6:* 1–12 (1944).

252 Malliani, A.; Ricordati, G.; Schwartz, P.J.: Nervous activity of afferent cardiac sympathetic fibers and atrial and ventricular endings. J. Physiol., Lond. *229:* 457–469 (1973).

253 Malliani, A.; Schwartz, P.J.; Aanchetti, A.: A sympathetic reflex elicited by experimental coronary occlusion. Am. J. Physiol. *217:* 703–709 (1969).

254 Manger, W.M.: Catecholamines and the heart; in Bourne, Hearts and heart-like organs. Physiology, vol. 2, pp. 161–262 (Academic Press, New York 1980).

255 Manger, W.M.; Gifford, R.W.: Pheochromocytoma (Springer, New York 1977).

256 Manger, W.M.; Estorff, I. von; Davis, S.; Chu, D.; Wakim, K.; Dufton, S.: Inadequacy of plasma catecholamines as an index of adrenergic activity. Fed. Proc. *34:* Abstr. 2853, p. 723 (1975).

256a Manger, W.M.; Hulse, M.C.; Forsyth, M.S.; Chute, R.N.; Brown, C.E.; Webb, K.; Sussman, R.; Warren, S.: Effect of pheochromocytoma and hypophysectomy on BP and catecholamines in NEDH rats. Hypertension, suppl. 4, pp. 200–207 (1982).

257 Manger, W.M.; Wakim, K.G.; Bollman, J.L.: Chemical quantitation of epinephrine and norepinephrine in plasma (Thomas, Springfield 1959).

258 Manning, J.W.: Intracranial mechanisms of regulation; in Randall, Neural regulation of the heart, pp. 189–209 (Oxford University Press, New York 1977).

259 Mark, A.L.; Mayer, H.E.; Schmid, P.G.; Heistad, D.D.; Abboud, F.M.: Adrenergic control of the peripheral circulation in cardiomyopathic hamsters with heart failure. Circulation Res. *33:* 74–81 (1973).

260 Maroko, P.R.; Bernstein, E.F.; Libby, P., DeLaria, G.A.; Covell, J.W.; Ross, J., Jr.; Braunwald, E.: Effects of intraaortic balloon counterpulsation on the severity of myocardial ischemic injury following acute coronary occlusion. Circulation *45:* 1150–1159 (1972).

261 Maroko, P.R.; Braunwald, E.: Effects of metabolic and pharacologic interventions on myocardial infarct size following coronary occlusion. Circulation *53:* suppl. I, 162–168 (1976).

262 Maroko, P.R.; Braunwald, E.; Covell, J.W.; Ross, J., Jr.: Factors influencing the severity of myocardial ischemia following experimental coronary occlusion (Abstract). Circulation *40:* suppl. III, p. 130 (1969).

263 Maroko, P.R.; Kjekshus, J.K.; Sobel, B.E.; Wantanabe, T.; Covell, J.W.; Ross, J., Jr.; Braunwald, E.: Factors influencing infarct size following coronary artery occlusion. Circulation *43:* 67–82 (1971).

264 Maroko, P.R.; Libby, P.; Covell, J.W.; Sobel, B.E.; Ross, J., Jr.; Braunwald, E.: Precordial S-T segment elevation mapping; an atraumatic method for assessing alterations in the extent of myocardial ischemic injury. Am. J. Cardiol. *29:* 223–230 (1972).

265 Mary-Rabine, L.; Hordof, A.J.; Bowman, F.O.; Malm, J.R.; Rosen, M.R.: Alpha and beta adrenergic effects on human atrial specialized conducting fibers. Circulation 57: 84–90 (1978).

266 Maseri, A.; L'Abbate, A.; Baroldi, G.; Chierchia, S.; Marzilli, M.; Ballestra, A.M.; Severi, S.; Parodi, O.; Biagini, A.; Distante, A.; Pesola, A.: Coronary vasospasm as a possible cause of myocardial infarction – a conclusion derived from the study of 'preinfarction angina'. New Engl. J. Med. 229: 1271–1277 (1978).

267 Mason, D.T.; Awan, N.A.; Joye, J.A.; Lee, G.; DeMaria, A.N.; Amsterdam, E.A.: Treatment of acute and chronic congestive heart failure by vasodilator-afterload reduction. Archs intern. Med. 140: 1577–1581 (1980).

268 Mason, D.T.; Braunwald, E.: Studies on digitalis. X. Effects of ouabain on forearm vascular resistance and venous tone in normal subjects and in patients with heart failure. J. clin. Invest. 43: 532–543 (1964).

269 Mason, J.W.; Stinson, E.B.; Harrison, D.C.: Autonomic nervous system and arrhythmias: studies in the transplanted denervated human heart. Cardiology 61: 75–87 (1976).

270 Massara, F.; Tripodina, A.; Rotunno, M.: Propranolol block of epinephrine induced by hypokaliaemia in man. Eur. J. Pharmacol. 10: 404–407 (1970).

271 Mathes, P.; Cowan, C.; Gudbjarnason, S.: Storage and metabolism of norepinephrine after experimental myocardial infaction. Am. J. Physiol. 220: 27–32 (1971).

272 Mathes, P.; Gudbjarnason, S.: Changes in norepinephrine stores in the canine heart following experimental myocardial infarction. Am. Heart J. 81: 211–219 (1971).

273 McCullagh, E.P.; Engel, W.J.: Pheochomocytoma with hypermetabolism. Report of two cases. Ann. Surg. 116: 61–75 (1942).

274 McDonald, L.; Baker, C.; Bray, C.; McDonald, A.; Restieaux, N.: Plasma catecholamines after cardiac infarction. Lancet ii: 1021–1023 (1969).

275 Medical Letter, The: in Abramowicz, Verapamil for arrhythmias, vol. 23, pp. 29–30 (The Medical Letter, New Rochelle 1981).

276 Meirson, F.Z.: The myocardium in hyperfunction, hypertrophy, and heart failure. Circulation Res. 25: suppl. II, p. 143 (1969).

277 Melville, K.I.; Blum, B.; Shister, H.E.; Silver, M.D.: Cardiac ischemic changes and arrhythmias induced by hypothalamic stimulation. Am. J. Cardiol. 12: 781–791 (1963).

278 Meyer, G.A.; Winter, D.L.: Spinal cord participation in the Cushing reflex in the dog. J. Neurosurg. 33: 662–675 (1970).

279 Miura, Y.: Plasma and tissue catecholamines in human cardiovascular disorders; in Mezey, Caldwell, Catecholamines and the heart; Int. Congr. Symp. ser., No. 8, pp. 59–68 (Royal Society of Medicine, London/Academic Press, London/Grune & Stratton, New York 1979).

280 Molzahn, M.; Dissman, T.; Gotzen, R.; Lohmann, F.W.: The effect of acute and chronic diminution of cardiac output on blood pressure in early hypertension. Hemodynamic alterations after β-receptor blockade. Klin. Wschr. 49: 476–484 (1971).

281 Molzahn, M.; Dissmann, T.; Halim, S.; Lohmann, F.W.; Oelkers, W.: Orthostatic changes of haemodynamics, renal function, plasma catecholamines and plasma renin concentration in normal and hypertensive man. Clin. Sci. 42: 209–222 (1972).

282 Moore, E.N.; Morse, H.T.; Price, H.L.: Cardiac arrhythmias produced by catecholamines in anesthetized dogs. Circulation Res. 15: 77–82 (1964).

283 Moskowitz, M.A.; Wurtman, R.J.: Catecholamines and neurologic diseases (first of two parts). New Engl. J. Med. *292:* 274–280 (1975).

284 Moskowitz, M.A.; Wurtman, R.J.: Catecholamines and neurologic diseases (second of two parts). New Engl. J. Med. *292:* 332–338 (1975).

285 Mudge, G.H., Jr.: Grossman, W.; Mills, R.M., Jr.; Lesch, M.; Braunwald, E.: Reflex increase in coronary vascular resistance in patients with ischemic heart disease. New Engl. J. Med. *295:* 1333–1337 (1976).

286 Mueller, H.S.: Preservation of ischemic myocardium in man. Cardiovasc. Rev. Rep. *2:* 33–40 (1981).

287 Mueller, H.S.; Ayres, S.M.; Conklin, E.F.; Giannelli, S., Jr.; Mazzara, J.T.; Grace, W.T.; Neal, T.F., Jr.: The effects of intra-aortic counterpulsation on cardiac performance and metabolism in shock associated with acute myocardial infarction. J. clin. Invest. *50:* 1885–1900 (1971).

288 Mueller, H.S.; Ayres, S.M.; Reglia, A.; Evans, R.G.: Propranolol in the treatment of acute myocardial infarction. Effect on myocardial oxygenation and hemodynamics. Circulation *49:* 1078–1087 (1974).

289 Multicenter International Study: Improvement in prognosis of myocardial infarction by long-term beta-adrenoreceptor blockade using practolol. Br. med. J. *iii:* 735–740 (1975).

290 Murray, R.H.; Luft, F.C.; Bloch, R.; Weyman, A.E.: Blood pressure responses to extremes of sodium intake in normal man. Proc. Soc. exp. Biol. Med. *159:* 432–436 (1978).

291 Nadeau, R.; De Champlain, J.; Gauthier, P.; Lavallée, M.; Peronnet, F.; Yamaguchi, N.: The catecholamines of the heart; in Mezey, Caldwell, Catecholamines and the heart. Int. Congr. Symp. ser., No. 8, pp. 1–11 (Royal Society of Medicine, London/Academic Press, London/Grune & Stratton, New York 1979).

292 Nagatsu, T.; Levitt, M.; Udenfriend, S.: Tyrosine hydroxylase – the initial step in norepinephrine biosynthesis. J. biol. Chem. *239:* 2910–2917 (1964).

293 Nayler, W.G.; Carson, V.: Effect of stellate ganglion stimulation on myocardial blood flow, oxygen consumption, and cardiac efficiency during beta-adrenoreceptor blockade. Cardiovasc. Res. *7:* 22–29 (1973).

294 Nayler, W.G.; Ferrari, R.; Williams, A.: Protective effect of pretreatment with verapamil, nifedipine and propranolol on mitochondrial function in the ischemic and reperfused myocardium. Am. J. Cardiol. *46:* 242–248 (1980).

295 Nelson, P.G.: Effect of heparin on serum free fatty acids, plasma catecholamines and the incidence of arrythmias following acute myocardial infarction. Br. med. J. *iii:* 735–737 (1970).

296 Nicholls, M.G.; Kiowski, W.; Zweifler, A.J.; Julius, S.; Schork, M.A.; Greenhouse, J.: Plasma norepinephrine variations with dietary sodium intake. Hypertension *2:* 29–32 (1980).

297 Norwegian Multicenter Study: Timolol-induced reduction in mortality and reinfarction in patients surviving acute myocardial infarction. New Engl. J. Med. *304:* 801–807 (1981).

298 Opie, L.H.; Nathan, D.; Lubbe, W.F.: Reviews: biochemical aspects of arrhythmogenesis and ventricular fibrillation. Am. J. Cardiol. *43:* 131–148 (1979).

299 Östman, I.; Sjostrand, H.O.; Swedin, G.: Cardiac noradrenaline turnover and urinary catecholamine excretion in trained and untrained rats during rest and exercise. Acta physiol. scand. *86:* 299–308 (1972).

300 Östman-Smith, I.: Adaptive changes in the sympathetic nervous system and some effector organs of the rat following long-term exercise or cold acclimation and the role of cardiac sympathetic nerves in the genesis of compensatory cardiac hypertrophy. Acta physiol. scand. *477:* suppl., pp. 1–118 (1979).

301 Otten, U.; Theonen, H.: Circadian rhythm of tyrosine-hydroxylase induction by short-term cold stress. Modulatory action of corticoids in newborn and adult rats. Proc. natn. Acad. Sci. USA *72:* 1415–1419 (1975).

302 Outschoorn, A.S.; Vogt, M.: The nature of cardiac sympathin in the dog. Br. J. Pharmacol. *7:* 319–324 (1952).

303 Paar, G.H.; Wellhöner, H.H.: The action of tetanus toxin on preganglionic sympathetic reflex discharges. Arch. Pharmacol. *276:* 437–445 (1973).

304 Pace, J.B.: Autonomic control of the coronary circulation; in Randall, Neural regulation of the heart, pp. 315–344 (Oxford University Press, New York 1977).

305 Pagano, V.T.; Incheosa, M.A., Jr.: Cardiomegaly produced by chronic beta-adrenergic stimulation in the rat: comparison with alpha-adrenergic effects. Life Sci. *21:* 619–624 (1977).

306 Palmer, G.C.; Spurgeon, H.A.; Priola, D.V.: Involvement of adenylate cyclase in mechanisms of denervation supersensitivity following surgical denervation of the dog heart. J. cyclic nucleotide Res. *1:* 89–95 (1975).

307 Parkes, D.: Bromocryptine. Adv. Drug Res. *12:* 247 (1976).

308 Pelkonen, R.; Pitkanen, E.: Unusual electrocardiographic changes in pheochromocytoma. Acta med. scand. *173:* 41–44 (1963).

309 Perrin, A.; Normand, J.; Mornex, R.; Froment, R.: Pheochromocytome et atherosclerose coronarrienne precoce. Rev. Atheroscler. *2:* 211–221 (1960).

310 Pincoffs, M.C.: A case of paroxysmal hypertension associated with suprarenal tumor. Trans. Ass. Am. Physns *44:* 295–299 (1929).

311 Pitt, B.; Elliot, E.C.; Gregg, D.E.: Adrenergic receptor activity in the coronary arteries of the unanesthetized dog. Circulation Res. *21:* 75–84 (1967).

312 Pitt, B.; Green, H.L.; Sugishita, Y.: Effect of beta-adrenergic receptor blockade on coronary haemodynamics in the resting unanesthetized dog. Cardiovasc. Res. *4:* 89–92 (1970).

313 Pool, P.E.; Covell, J.W.; Levitt, M.; Gibb, J.; Braunwald, E.: Reduction in cardiac tyrosine hydroxylase activity in experimental congestive heart failure – its role in the depletion of cardiac norepinephrine stores. Circulation Res. *20:* 349–353 (1967).

314 Potter, L.T.; Axelrod, J.: Properties of norepinephrine storage particles of the rat heart. J. Pharmac. exp. Ther. *442:* 299–305 (1963).

315 Potter, L.T.; Cooper, T.; Willman, V.L.; Wolfe, D.E.: Synthesis, binding, release and metabolism of norepinephrine in normal and transplanted dog hearts. Circulation Res. *16:* 468–481 (1965).

316 Priola, D.V.: Individual chamber sensitivity to norepinephrine after unilateral cardiac denervation. Am. J. Physiol. *216:* 604–614 (1969).

317 Pullman, T.N.; McClure, W.W.: The effect of *L*-noradrenaline on electrolyte excretion in normal man. J. Lab. clin. Med. *39:* 711–719 (1952).

318 Raab, W.: Hormonal and neurogenic cardiovascular disorders; endocrine and neuro-endocrine factors in pathogenesis and treatment (Williams & Wilkins, Baltimore 1953).

319 Raab, W.; Gigee, W.: Norepinephrine and epinephrine content of normal and diseased human hearts. Circulation *11:* 593–603 (1955).

320 Rabinowitz, M.; Zak, R.: Biochemical and cellular changes in cardiac hypertrophy. A. Rev. Med. *23:* 245–262 (1972).

321 Radtke, W.E.; Kazmier, F.J.; Rutherford, B.D.; Sheps, S.G.: Cardiovascular complications of pheocromocytoma crisis. Am. J. Cardiol. *35:* 701–705 (1975).

322 Randall, W.C. (ed.): Neural regulation of the heart (Oxford University Press, New York 1977).

323 Randall, W.C.; Armour, J.A.: Gross and microscopic anatomy of the cardiac innervation; in Randall, Neural regulation of the heart, pp. 13–41 (Oxford University Press, New York 1977).

324 Randall, W.C.; Armour, J.A.; Geis, W.P.; Lippincott, D.B.: Regional cardiac distribution of the sympathetic nerves. Fed. Proc. *31:* 1199–1208 (1972).

325 Randall, W.L.; Euler, D.E.; Jacobs, H.K.; Wehrmacher, W.; Kaye, M.P.; Hageman, G.R.: Autonomic neural control of cardiac rhythm: the role of autonomic imbalance in the genesis of cardiac dysrhythmia. Cardiology *61:* 20–36 (1976).

326 Reder, R.F.; Rosen, M.R.: The role of the sympathetic nervous system in sudden cardiac death. Drug Ther. *3:* 43–55 (1978).

327 Roberts, J.; Kelliher, G.: Effect of practolol and sotalol on adrenergic nervous activity. Fed. Proc. *32:* 780 (1970).

328 Robertson, D.H.; Johnson, G.A.; Brilis, G.M.; Hill, R.E.; Watson, J.T.; Oates, J.A.: Salt restriction increases serum catecholamines and urinary metanephrine excretion (Abstract). Fed. Proc. *36:* 956 (1977).

329 Rogoff, J.M.; Quashnock, J.M.; Nixon, E.N.; Rosenberg, A.W.: Adrenal function and blood electrolytes. Proc. Soc. exp. Biol. Med. *73:* 163–169 (1950).

330 Romanoff, M.S.; Keusch, G.; Campese, V.M.; Wang, M.S.; Friedler, R.M.; Weidmann, P.; Massry, S.G.: Effect of sodium intake on plasma catecholamines in normal subjects. J. clin. Endocr. Metab. *48:* 26–31 (1978).

331 Rona, G.; Chappel, C.I.; Balazs, T.; Gaudry, R.: Infarct-like myocardial lesion and other toxic manifestations produced by isoproterenol in the rat. Archs Path. *67:* 443–455 (1959).

332 Rose, A.G.: Catecholamine induced myocardial damage associated with pheochromocytoma and tetanus. S. Afr. Med. J. *48:* 1285–1289 (1974).

333 Rosen, M.R.; Hordof, A.J.; Ilvento, J.P.; Danilo, P., Jr.: Effect of adrenergic amines on electrophysiological properties and automaticity of neonatal and adult canine Purkinje fibers. Evidence for α- and β-adrenergic actions. Circulation Res. *40:* 390–400 (1977).

334 Rothberger, J.; Winterberger, H.: Über die Beziehungen der Herznerven zur Form des Electrokardiogramms. Pflügers Arch. ges. Physiol. *135:* 506 (1910).

335 Rubio, R.; Berne, R.M.: Regulation of coronary blood flow. Prog. cardiovasc. Dis. *18:* 105–122 (1975).

336 Russell, D.C.; Oliver, M.F.: Ventricular refractoriness during acute myocardial ischaemia and its relationship to ventricular fibrillation. Cardiovasc. Res. *12:* 221–227 (1978).

337 Russell, R.A.; Crafoord, J.; Harris, A.S.: Changes in myocardial composition after coronary artery ligation. Am. J. Physiol. *200:* 995–998 (1961).

338 Rutenberg, H.L.; Spann, J.F., Jr.: Alterations of cardiac sympathetic neurotransmitter activity in congestive heart failure; in Mason, Congestive heart failure: mechanisms, evaluation, and treatment, pp. 85–95 (Yorke Medical Books, New York 1976).

339 Saavedra, J.M.; Grobecker, H.; Axelrod, J.: Adrenaline forming enzyme in the brain stem: Elevation in genetic and experimental hypertension. Science 191: 483–484 (1976).

340 Safar, M.E.; Weiss, Y.A.; Leverson, J.A.; London, G.M.; Milliez, P.L.: Hemodynamic study of 85 patients with borderline hypertension. Am. J. Cardiol. 31: 315–319 (1973).

341 Saint-Pierre, A.; Lejosne, C.; Perrin, A.: Aspects electrocardiographiques des pheochromocytomes. Coeur Méd. interne 13: 59–73 (1974).

342 Saint-Pierre, A.; Perrin, A.; Mornex, R.; Pouzeratte, J.-P.: Les troubles du rythme cardiaque dus aux pheochromocytomes (et forme rythmique pure). Coeur Méd. interne 9: 3–11 (1970).

343 Sannerstedt, R.: Hemodynamic findings at rest and during exercise in mild central hypertension. Am. J. med. Sci. 258: 70–79 (1969).

344 Sarnoff, S.J.; Gilmore, J.P.; Brockman, S.K.; Mitchell, J.H.; Linden, R.J.: Regulation of ventricular contraction by the carotid sinus: Its effect on atrial and ventricular dynamics. Circulation Res. 8: 1123–1136 (1960).

345 Sarnoff, S.J.; Mitchell, J.H.: The control of the function of the heart; in Hamilton, Dow, Handbook of physiology, sect. 2, vol. I, pp. 489–532 (Am. Physiological Society, Washington 1962).

346 Sayer, W.J.; Moser, M.; Mattingly, T.W.: Pheochromocytoma and the abnormal electrocardiogram. Am. Heart. J. 48: 42–53 (1954).

347 Schaal, S.F.; Wallace, A.G.; Sealy, W.C.: Protective influence of cardiac denervation against arrhythmias of myocardial infarction. Cardiovasc. Res. 3: 241–244 (1969).

348 Schuelke, D.M.; Mark, A.L.; Schmid, P.G.; Eckstein, J.W.: Coronary vasodilatation produced by dopamine after adrenergic blockade. J. Pharmac. exp. Ther. 176: 320–327 (1971).

349 Schwartz, P.J.; Malliani, A.: Electrical alternation of the T-waves: Clinical and experimental evidence of its' relationship with the sympathetic nervous system and with the long Q-T syndrome. Am. Heart J. 89: 45–50 (1975).

350 Scott, N.A.; DeSilva, R.A.; Lown, B.; Wurtman, R.J.: Tyrosine administration decreases vulnerability to ventricular fibrillation in the normal canine heart. Science 211: 727–729 (1981).

351 Sen, S.K.; Khairallah, P.A.; Tarazi, R.C.; Bumpus, F.M.: Biochemical changes associated with development and reversal of cardiac hypertrophy in spontaneously hypertensive rats. Cardiovasc. Res. 10: 254–261 (1976).

352 Sen, S.; Tarazi, R.C.; Bumpus, F.M.: Cardiac hypertrophy and antihypertensive therapy. Cardiovasc. Res. 11: 427–433 (1977).

353 Sen, S.; Tarazi, R.C.; Bumpus, F.M.: Cardiac effects of angiotensin antagonists in normotensive rats. Clin. Sci. mol. Med. 56: 439–443 (1979).

354 Shapiro, A.P.; Baker, H.M.; Hoffman, M.S.; Pharmacologic and physiologic studies of a case of pheochromocytoma. Am. J. Med. 10: 115–130 (1951).

355 Shaver, J.A.; Leon, D.F.; Graw, S., III; Leonard, J.J.; Bahnson, H.T.: Haemodynamic observations after cardiac transplantation. New Engl. J. Med. 281: 822–827 (1969).

356 Shindler, R.; Harakal, C.; Sevy, R.W.: Catecholamine content of the sino-atrial node and common right atrial tissue. Proc. Soc. exp. Biol. Med. *128:* 789–800 (1968).

357 Siegel, J.H.; Gilmore, J.P.; Sarnoff, S.J.: Myocardial extraction and production of catecholamines. Circulation Res. *9:* 1336–1350 (1961).

358 Simeone, F.A.; Sarnoff, S.J.: The effect of dibenamine on autonomic stimulation. Surgery, St Louis *22:* 391–401 (1947).

359 Singh, B.N.; Burnam, M.H.: The role of beta-adrenergic blocking drugs in early myocardial infarction. Cardiovasc. Rev. Rep. *1:* 281–287 (1980).

360 Sjöstrand, T.: After-potentials in the elctrocardiogram. Acta physiol. scand. *24:* 247–260 (1951).

361 Skelton, R.B.; Gergely, N.; Manning, G.W.; Coles, J.C.: Mortality studies in experimental coronary occlusion. J. thorac. cardiovasc. Surg. *44:* 90–96 (1962).

362 Sleight, P.: Beta-adrenergic blockade after myocardial infarction. New Engl. J. Med. *304:* 837–838 (1981).

363 Snyder, S.H.: Receptors, neurotransmitters and drug responses. New Engl. J. Med. *300:* 465–472 (1979).

364 Sode, J.; Getzen, L.C.; Osborne, D.P.: Cardiac arrhythmias and cardiomyopathy associated with pheochromocytomas. Am. J. Surg. *114:* 927–931 (1967).

365 Sole, M.J.; Kamble, A.B.; Hussain, M.N.: A possible change in the rate-limiting step for cardiac norepinephrine synthesis in the cardiomyopathic Syrian hamster. Circulation Res. *41:* 814–817 (1977).

366 Spann, J.F., Jr.; Buccino, R.A.; Sonnenblick, E.H.: Production of right ventricular hypertrophy with and without congestive heart failure in the cat. Proc. Soc. exp. Biol. Med. *125:* 522–524 (1967).

367 Spann, J.F., Jr.; Buccino, R.A.; Sonnenblick, E.H.; Braunwald, E.: Contractile state of cardiac muscle obtained from cats with experimentally produced ventricular hypertrophy and heart failure. Circulation Res. *21:* 341–354 (1967).

368 Spann, J.F., Jr.; Chidsey, C.A.; Braunwald, E.: Reduction of cardiac stores of norepinephrine in experimental heart failure. Science *145:* 1439–1441 (1964).

369 Spann, J.F., Jr.; Chidsey, C.A.; Pool, P.E.; Braunwald, E.: Mechanism of norepinephrine depletion in experimental heart failure produced by aortic constriction in the guinea pig. Circulation Res. *17:* 312–321 (1965).

370 Spector, S.; Sjoerdsma, A.; Zaltzman-Nirenberg, P.; Levitt, M.; Udenfriend, S.: Conversion of tyrosine-C^{14} to norepinephrine by the isolated perfused guinea pig heart. Science *139:* 1299–1301 (1963).

371 Spotnitz, H.M.; Beach, P.M.; Brigman, D.; Truccone, N.; Parodi, E.N.; Manger, W.M.; Malm, J.R.: Oxygen conservation by propranolol and intra-aortic balloon pumping in the normal canine heart. J. surg. Res. *22:* 453–462 (1977).

372 Spurgeon, W.A.; Priola, D.V.; Montoya, P.; Weiss, G.K.; Alter, W.A.: Catecholamines associated with conductile and contractile myocardium of normal and denervated dog hearts. J. Pharmac. exp Ther. *190:* 466–471 (1974).

373 Stanton, H.C.; Brenner, G.; Mayfield, E.D.: Studies on isoproterenol-induced cardiomegaly in rats. Am. Heart J. *77:* 72–80 (1969).

374 Stanton, H.C.; Vick, R.L.: Cholinergic and adrenergic influences on right ventricular myocardial contractility in the dog. Archs int. Pharmacodyn. Ther. *176:* 233–248 (1968).

375 Starke, K.; Taube, H.D.; Borowski, E.: Presynaptic receptor systems in catecholaminergic transmission. Biochem. Pharmacol. *26:* 259–268 (1977).

376 Stephen, S.A.: Unwanted effects of propranolol. Am. J. Cardiol. *18:* 463–472 (1966).
377 Stinson, E.B.; Griepp, R.B.; Clark, D.A.; Dong, E.; Shumway, N.E.: Cardiac transplantation in man. VIII. Survival and function. J. thorac. cardiovasc. Surg. *60:* 303–319 (1970).
378 Stinson, E.B.; Griepp, R.B.; Schroeder, J.S.; Dong, E., Jr.; Shumway, N.E.: Hemodynamic observations one and two years after cardiac transplantation in man. Circulation *45:* 1183–1194 (1972).
379 Stjärne, L.; Brundin, J.: Dual adrenoceptor-mediated control of noradrenaline secretion from human vasoconstrictor nerves: facilitation by β-receptors and inhibition by α-receptors. Acta physiol. scand. *94:* 139–141 (1975).
380 Sutherland, E.W.; Robison, G.A.; Butcher, R.W.: Some aspects of the biological role of adenosine 3′, 5′-monophosphate (cyclic AMP). Circulation *37:* 279–306 (1968).
381 Szabó, J.; Nosztray, K.; Csáky, L.; Szegi, J.: Prevention of isoproterenol-induced cardiac hypertrophy by beta-blocking agents in the rat. Acta physiol. hung. *48:* 79–86 (1976).
382 Szentiványi, M.; Pace, J.P.; Wechsler, J.S.; Randall, W.C.: Localized myocardial responses to stimulation of cardiac sympathetic nerves. Circulation Res. *21:* 691–702 (1967).
383 Tada, M.; Kirchberger, M.A.; Iorio, J.-A.; Katz, A.M.: Control of cardiac sarcolemmal adenylate cyclase and sodium, potassium-activated adenosinetriphosphatase activities. Circulation Res. *36:* 8–17 (1975).
384 Tarazi, R.C.; Sen, S.: Catecholamines and cardiac hypertrophy; in Mezey, Caldwell, Catecholamines and the heart. Int. Congr. Symp. ser., No. 8, pp. 47–57 (Royal Society of Medicine, London/Academic Press, London/Grune & Stratton, New York 1979).
385 Tenzer, C.: Quelques aspects electrocardiographiques d'un pheochromocytome. Acta cardiol. *9:* 532–541 (1954).
386 Thomas, J.A.; Marks, B.H.: Plasma norepinephrine in congestive heart failure. Am. J. Cardiol. *41:* 233–243 (1978).
387 Tomoike, H.; Franklin, D.; McKown, D.; Kemper, W.S.; Guberer, M.; Ross, J., Jr.: Regional myocardial dysfunction and hemodynamic abnormalities during strenuous exercise in dogs with limited coronary flow. Circulation Res. *42:* 487–496 (1978).
388 Tomoike, H.; Ross, J., Jr.; Franklin, D.; Grozatier, B.; McKown, D.; Kemper, W.S.: Improvement by propranolol of regional myocardial dysfunction and abnormal coronary flow pattern in conscious dogs with coronary narrowing. Am. J. Cardiol. *41:* 689–696 (1978).
389 Tsien, R.W.; Giles, W.R.; Greengard, P.: Cyclic AMP mediates the action of epinephrine on the action potential plateau of cardiac Purkinje fibers. Nature new Biol. *240:* 181–183 (1972).
390 Ueda, H.; Yanai, Y.; Marav, S.; Haruine, K.; Mashima, S.; Kuroiwa, A.; Sugimoto, T.; Shimomura, K.: Electrocardiographic and vetorcardiographic changes produced by electrical stimulation of the cardiac nerves. Jap. Heart J. *5:* 359–372 (1964).
391 Ulmer, R.H.; Randall, W.C.: Atrioventricular pressures and their relationships during stellate stimulation. Am. J. Physiol. *201:* 134–138 (1961).
392 Urthaler, F.; James, T.N.: Effect of tetrodotoxin on A-V conduction and A-V junctional rhythm. Am. J. Physiol. *224:* 1155–1161 (1973).

393 Urthaler, F.; James, T.N.: Cholinergic and adrenergic control of the sinus node and AV junction; in Randall, Neural regulation of the heart, pp. 247–288 (Oxford University Press, New York 1977).

394 Urthaler, F.; Katholi, C.R.; Macy, J., Jr.; James, T.N.: Mathematical relationships between automaticity of the sinus node and AV junction. Am. Heart J. *86:* 189–195 (1973).

395 Urthaler, F.; Katholi, C.R.; Macy, J., Jr.; James, T.N.: Electro-physiological and mathematical characteristics of the escape rhythm during complete AV block. Cardiovasc. Res. *8:* 173–186 (1974).

396 Van Citters, R.L.; Smith, O.A.; Ruttenberg, H.D.: Subthalamically induced paroxysmal ventricular tachycardia after complete heart block. Am. J. Physiol. *211:* 293–300 (1966).

397 Vander, A.J.: Effect of catecholamines and the renal nerves on renin secretion in anesthetized dogs. Am. J. Physiol. *209:* 659–662 (1965).

398 Vanderbeek, R.B.; Ebert, P.A.: Potassium release in the denervated heart. Am. J. Physiol. *218:* 803–806 (1970).

399 Vane, J.R.: The second Gaddum Memorial lecture. The release and fate of vasoactive hormones in the circulation. Br. J. Pharmacol. *35:* 209–242 (1969).

400 Van Vliet, P.D.; Burchell, H.B.; Titus, J.L.: Focal myocarditis associated with pheochromocytoma. New Engl. J. Med. *274:* 1102–1108 (1966).

401 Vatner, S.F.; Braunwald, E.: Cardiovascular control mechanisms in the conscious state. New Engl. J. Med. *293:* 970–976 (1975).

402 Vatner, S.F.; Franklin, D.; Van Citters, R.L.; Braunwald, E.: Effect of carotid sinus nerve stimulation on the coronary circulation of the conscious dog. Circulation Res. *27:* 11–21 (1970).

403 Vatner, S.F.; Higgins, C.B.; Braunwald, E.: Effects of norepinephrine on coronary circulation and left ventricular dynamics in the conscious dog. Circulation Res. *34:* 812–823 (1975).

404 Vatner, S.F.; Millard, R.W.; Higgins, C.B.: Coronary and myocardial effects of dopamine in the conscious dog: Parasympathetic augmentation of pressor and myotropic actions. J. Pharmac. exp. Ther. *187:* 280–295 (1973).

405 Vatner, S.F.; McRitchie, R.J.: Responses of coronary smooth muscle to catecholamines. Circulation Res. *39:* 664–673 (1975).

406 Vaughan Williams, E.M.; Raine, A.E.G.; Cabrera, A.A.; Whyte, J.M.: The effects of prolonged β-adrenoceptor blockade on heart weight and cardiac intracellular potentials in rabbits. Cardiovasc. Res. *9:* 579–592 (1975).

407 Vendsalu, A.: Studies on adrenaline and noradrenaline in human plasma. Acta physiol. scand. *49:* suppl. 173, pp. 1–123 (1960).

408 Verde, G.; Oppizzi, G.; Colussi, G.; Cremascoli, G.; Botalla, L.; Müller, E.E.; Silvestrini, F.; Chiodini, P.G.; Liuzzi, A.: Effect of dopamine infusion on plasma levels of growth hormone in normal subjects and in acromegalic patients. Clin. Endocrinol. *5:* 419–423 (1976).

409 Viveros, O.H.; Arqueros, L.; Kirshner, N.: Release of catecholamines and dopamine oxidase from the adrenal medulla. Life Sci. *7:* 609–618 (1968).

410 Vogel, J.H.K.; Chidsey, C.A.: Cardiac adrenergic activity in experimental heart failure assessed with beta-receptor blockade. Am. J. Cardiol. *24:* 198–208 (1969).

411 Watkins, D.B.: Pheochromocytoma: a review of the literature. J. chron. Dis. *6:* 510–527 (1957).

412 Weber, K.T.; Kinasewitz, G.T.; West, J.S.; Janicki, J.S.; Reichek, N.; Fishman, A.P.: Long-term vasodilator therapy with Trimazosin in chronic cardiac failure. New Engl. J. Med. *303:* 242–250 (1980).

413 Weil-Malherbe, H.; Axelrod, J.; Tomchick, R.: Blood-brain barrier for adrenaline. Science *129:* 1226–1227 (1959).

414 Weinshilboum, R.M.; Thoa, N.B.; Johnson, D.G.; Kopin, I.J.; Axelrod, J.:Proportional release of norepinephrine and dopamine-β-hydroxylase from sympathetic nerves. Science *174:* 1349–1351 (1971).

415 Werning, C.; Fischer, N.; Kaip, E.; Stiel, D.; Trubenstein, G.K.; Vetter, H.: Increased renin stimulation after orthostasis in labile or borderline hypertension. Dt. med. Wschr. *97:* 1038–1039 (1972).

416 West, G.B.; Shepard, D.M.; Hunter, R.B.: Adrenaline and noradrenaline concentration in adrenal glands at different ages and in some diseases. Lancet *ii:* 966–969 (1951).

417 West, G.B.; Shepherd, D.M.; Hunter, R.B.; MacGregor, A.R.: The function of the organs of Zuckerkandl. Clin. Sci. *12:* 317–325 (1953).

418 Westfall, T.C.: Local regulation of adrenergic neurotransmission. Physiol. Rev. *57:* 659–728 (1977).

419 Whitby, L.G.; Axelrod, J.; Weil-Malherbe, H.: The fate of H³-norepinephrine in animals. J. Pharmac. exp. Ther. *132:* 193–201 (1961).

420 Wilhelmsson, C.; Vedin, J.A.; Wilhelmsen, L.; Tibblin, G.; Werkö, L.: Reduction of sudden deaths after myocardial infarction by treatment with alprenolol. Lancet *ii:* 1157–1160 (1974).

421 Williams, L.T.; Lefkowitz, R.J.: Receptor binding studies in adrenergic pharmacology (Raven Press, New York 1978).

422 Williams, N.J. (ed.): Therapeutic advances in medicine: nifedipine (symposium). Br. J. clin. Pract. suppl. 8 (1980).

423 Williams, T.H.: Electron microscopic evidence for an autonomic interneuron. Nature, Lond. *214:* 309–310 (1967).

424 Witham, A.C.; Fleming, J.W.: The effect of epinephrine on the pulmonary circulation in man. J. clin. Invest. *30:* 707–717 (1951).

425 Witzke, D.J.; Kaye, M.P.: Myocardial ultrastructural changes induced by administration of nerve growth factor. Surg. Forum *27:* 295–297 (1976).

426 Wurster, R.D.: Spinal sympathetic control of the heart; in Randall, Neural regulation of the heart, pp. 213–246 (Oxford University Press, New York 1977).

427 Wurster, R.D.; Randall, W.C.: Cardiovascular responses to bladder distension in patients with spinal transection. Am. J. Physiol. *228:* 1288–1292 (1975).

428 Wurtman, R.J.: Catecholamines (Little, Brown, Boston 1966).

429 Wurtman, R.J.; Kopin, I.J.; Axelrod, J.: Thyroid function and the cardiac disposition of catecholamines. Endocrinology *73:* 63–74 (1963).

430 Yamaguchi, N.; De Champlain, J.; Nadeau, R.: Correlation between the response of the heart to sympathetic stimulation and the release of endogenous catecholamines into the coronary sinus of the dog. Circulation Res. *36:* 662–668 (1975).

431 Yamaguchi, N.; De Champlain, J.; Nadeau, R.A.: Regulation of norepinephrine release from cardiac sympathetic fibers in the dog by pre-synaptic α- and β-receptors. Circulation Res. *41:* 108–117 (1977).

432 Yamaguchi, A.: Ultrastructure of the innervation of the mammalian heart; in
Challice, Virágh, The ultrastructure of the mammalian heart, pp. 127–178 (Academic
Press, New York 1973).

433 Yamori, Y.: Organ differences of catecholamine metabolism in spontaneously hypertensive rats; in Okamoto, Spontaneous hypertension, pp. 59–61 (Igaku Shoin, Tokyo
1972).

434 Yamori, Y.: Neural and non-neural mechanisms in spontaneous hypertension. Clin.
Sci. mol. Med. *51:* 4315–4345 (1976).

435 Yankopoulos, N.A.; Montero, A.C.; Curd, W.G., Jr.; Kahil, M.E.; Condon, R.E.:
Observations on myocardial function during chronic catecholamine oversecretion.
Chest *66:* 585–587 (1974).

436 Yanowitz, F.; Preston, J.B.; Abildskov, J.A.: Functional distribution of right and
left stellate innervation to the ventricles. Production of neurogenic electrocardiographic changes by unilateral alteration of sympathetic tone. Circulation Res. *18:*
416–428 (1966).

437 Yasue, H.; Touyama, M.; Kato, H.; Tanaka, S.; Akiyama, F.: Prinzmetal's variant
form of angina as a manifestation of alpha-adrenergic receptor-mediated coronary
artery spasm: documentation by coronary arteriography. Am. Heart J. *91:* 148–155
(1976).

438 Yates, J.C.; Beamish, R.E.; Dhalla, N.S.: Ventricular dysfunction and necrosis
produced by adrenochrome metabolite of epinephrine: relation to pathogenesis of
catecholamine cardiomyopathy. Am. Heart J. *102:* 210–221 (1981).

439 Zelis, R.; Longhurst, J.; Capone, R.J.; Lee, G.: Peripheral circulatory control
mechanisms in congestive heart failure. Am. J. Cardiol. *32:* 481–490 (1973).

440 Zuberbuhler, R.C.; Bohr, D.F.: Responses of coronary smooth muscle to catecholamines. Circulation Res. *16:* 431–440 (1965).

Subject Index

Acetylcholine
catecholamine secretion in response to
16, 44
cyclic AMP 47
cyclic GMP 47
electrophysiological properties of heart,
effect 69
pacemaker response to vagal stimulation, effect 43, 44
Adenosine triphosphate storage 8
Adrenal gland
catecholamine biosynthesis 15
norepinephrine and epinephrine 15
Adrenergic nerves
hypertrophied heart 98
norepinephrine biosynthesis, storage
and release from 5–9
Adrenergic neurotransmission
agonists, effect on norepinephrine
release 9, 10
blockers, inhibition of norepinephrine
release 10, see also individual agents
normal and hypertrophied hearts
104
Adrenergic receptors
alpha
classification 17
desensitization to epinephrine 22
representative compounds 19
alpha and beta in coronary arteries 57
beta
blockade effect 27, 55
classification 17
representative compounds 18
desensitization and resensitization 21
dopaminergic 18
hormonal, genetic and environmental
influences 21

interaction with catecholamines 21
mechanism 22
myocardial 24–26
pharmacologic differentiation of
alpha and beta 19
Agonists
beta-adrenergic 19
desensitization and resensitization 22
exposure of receptors 21
AMP, cyclic
adrenergic response 21–23
intracellular levels 47
myocardial, vagal effect 47
Angina 48
coronary artery denervation 86
preinfarction 84
Prinzmetal's 84
spontaneous rest 84
Antagonists, beta-adrenergic 19
exposure of receptors 21
Antihypertensive drugs in cardiac hypertrophy 100
Aorta
ascending, blood flow, sympathetic
influence 37, 38
catecholamine plasma levels
after sympathetic nerve stimulation
37, 38
at rest and during exercise 35
norepinephrine and epinephrine 35
pressure, sympathetic influences 39
Arrhythmias
adrenergic activity, effects 107
autonomic nervous system effects 71
beta-adrenergic blockade 82
biochemical aspects 79
cardiac transplants 74
genesis 63, 78